SpringerBriefs in Applied Sciences and Technology

For further volumes:
http://www.springer.com/series/8884

Nirwan Ansari • Jingjing Zhang

Media Access Control and Resource Allocation

For Next Generation Passive Optical Networks

Nirwan Ansari
New Jersey Institute
    of Technology
Newark, NJ, USA

Jingjing Zhang
Arista Networks
Santa Clara, CA, USA

ISSN 2191-530X   ISSN 2191-5318 (electronic)
ISBN 978-1-4614-3938-7   ISBN 978-1-4614-3939-4 (eBook)
DOI 10.1007/978-1-4614-3939-4
Springer New York Heidelberg Dordrecht London

Library of Congress Control Number: 2012954389

Printed on acid-free paper

Springer is part of Springer Science+Business Media (www.springer.com)

To my wife, Jenny, for her unwavering support...

Nirwan Ansari

To my parents, brother, and husband for always being there for me.

Jingjing Zhang

Foreword

The success of the fiber network deployment in the 1990s fueled the explosive growth of the Internet for two decades. There is an ongoing joke among us the fiber optics researchers that we did such a great job on the research and development of fiber communication and at such a fast pace in the 1980s and 1990s that we finished all that we were expected to do in two decades and we burned ourselves out of a lucrative job market with our successes. That thesis seems to be confirmed by the recent job market in this area and the joke does not seem to be so funny anymore, at least to us in the field and particularly our graduate students. However, I submit that thesis is entirely wrong. The best research and development of fiber networks is still ahead of us. Let me explain.

The deployment of fiber communications in the first two decades was so rapid that uncovering the vast capability of the bandwidth of fiber was the first priority and low-hanging fruit. Almost any fiber system, optimum or not, was much better than the technology it replaced: copper. The major emphasis was on deployment not performance optimization. The computer science and network crowd superimposed the Internet architecture on the fiber substrate and the combination unleashed the broadband explosion. Most of the optical research community was coerced by the upper network layer community into making the speed faster, lower powered, and cheaper and leave the rest of the architecture to the upper layer folks to create. I always find that notions totally confining and not gratifying. I always believed and maintained that a network architecture that is created with the optical hardware properties and limitations in mind together with the right upper layer designs will yield much superior performance. Throughout the years there were a few of us who have conducted our research across all the network layers at once. The first author has been one of my very few fellow colleagues who has always been performing his research this way through feast and famine. This book is an excellent manifestation of this research philosophy and I laud the effort of the authors in committing the contents of this book in writing.

Though there are still lots of architectural works to be done in the long haul network, the bottleneck to network performance these days is in the access network. The optical access network deployed thus far has been rather unimaginative, using fiber almost as a high speed replacement of copper cable. This book opens up the possibility of using the interactions across the physical layer and the upper layers in designing a better media access control protocol and resource allocation algorithm. The book has a balanced combination of optical technology and architecture in the physical and MAC layer and MAC protocol and resource allocation algorithm and performance analysis in the upper layers. While the verdict is still out on what the future access network architecture will look like, the intellectual discipline as given in this book is a template for the pathway for a systematic approach to optical network research and development. The academic community will benefit from its broad treatment across both technology and architecture. The industry engineer will benefit from the mind-broadening approach to development and come to the realization that the development of the physical layer of the fiber network has been too myopic and confining in the past. Hopefully, both the academic and industrial communities will proceed with the future research and development of fiber networks using a broader and much more balanced approach as exemplified by this book.

Boston, MA   Vincent W.S. Chan
                                                               Joan and Irwin Jacobs Professor
                                                                              EECS, MIT

Preface

The Full Service Access Network (FSAN) working group, established by major telecommunications service providers and equipment vendors in 1995, has been advancing specifications into appropriate standard bodies, primarily ITU-T and IEEE, through which many standards have been defined and driven into passive optical network (PON) services and products. PON, which exploits the potential capacities of optical fibers, has also been identified as a promising future-proof access network technology to meet rapidly increasing traffic demands effectuated by the popularization of Internet and the sprouting of bandwidth-demanding applications. Tremendous amount of research and development efforts have since been made to advance PON.

This book is intended to provide a quick technical briefing on the state of the art of PON with respect to, in particular, media access control (MAC) and resource management. It consists of nine chapters: Chap. 1 provides the landscape of existing broadband access technologies; Chap. 2 traces the evolution of PON architectures from APON to OFDM PON; Chaps. 3–5 cover MAC and resource allocation in GPON, EPON, and WDM PON, respectively; Chaps. 6–8 cover recently zealously pursued topics, namely, OFDM PON, hybrid optical and wireless access, and green PON; and Chap. 9 presents the concluding remarks and points to future research and development endeavors. While Chaps. 1 and 2 provide rudimentary overviews, the material of the book is structured in a modular fashion, with each chapter reasonably independent of each other. Individual chapters can be perused in an arbitrary order to the liking and interest of each reader, and they can also be incorporated as part of a larger, more comprehensive course. The first author has adopted some material presented in this book for his graduate courses, ECE 639 Principles of Broadband Networks and ECE 788 Advanced Topics in Broadband Networks. The book may also be used as a reference for practicing networking engineers and researchers.

Newark, NJ   Nirwan Ansari
Santa Clara, CA   Jingjing Zhang

Contents

Acronyms

ADSL	Asymmetric DSL
AF	Assured forwarding
AGC	Automatic gain control
APON	ATM PON
ARPU	Average avenue per user
ATM	Asynchronous transfer mode
AWG	Arrayed waveguide grating
BBU	Base band unit
BE	Best effort
BPL	Broadband over PowerLine
BPON	Broadband PON
BSC	Base station controller
BWA	Broadband wireless access
BWmap	Bandwidth map
CAPEX	Capital expenditure
CAGR	Compound annual growth rate
CDR	Clock data recovery
DBA	Dynamic bandwidth allocation
DFB LD	Distributed feedback laser diode
DSL	Digital subscriber line
DSLAM	Digital subscriber line access multiplexer
EDF	Earliest-deadline-first
EF	Expedited forwarding
EFM	Ethernet in the first mile
ELAN	Ethernet LAN
EPON	Ethernet PON
E-Tree	Ethernet tree
EVC	Ethernet virtual connection
EV-DO	Evolution data optimized
FBA	Fixed bandwidth allocation
FCAPS	Fault, configuration, accounting, performance, security

FDMA	Frequency division multiple access
FEC	Forward error correction
FSAN	Full service access network
FTTB	Fiber-to-the-building
FTTC	Fiber-to-the-curb or cabinet
FTTD	Fiber-to-the-desk
FTTH	Fiber-to-the-home
FTTN	Fiber-to-the-node
FTTP	Fiber-to-the-premise
GEM	GPON encapsulation method
GEO	Geostationary Earth orbit
GPON	Gigabit PON
GTC	GPON transmission convergence
HDSL	High-bit-rate DSL
HFC	Hybrid fiber coaxial cable
HSPA	High speed packet access
ICT	Information and communication technology
IDSL	ISDN DSL
IPACT	Interleaved polling with adaptive cycle time
IPTV	Internet protocol television
ISDN	Integrated services digital network
LD	Laser diode
LEO	Low earth orbit
LFJ	Least flexible job
LFM	Least flexible machine
LLID	Logical link identification
LOS	Line of sight
LPT	Longest processing time
LTE	Long term evolution
MAC	Media access control
MAI	Multiple access interference
MEF	Metro ethernet forum
MEMS	Micro-electro-mechanical system
MIB	Management information base
MILP	Mixed integer linear programming
MPCP	MultiPoint control protocol
MTW	Maximum transmission window
NG-PON	Next-generation PON
NLOS	Non-LOS
NNI	Network network interface
NPE/PPE	Network/packet processing engine
OAM&P	Operation administration, maintenance, and provisioning
OCDMA	Optical code division multiple access
ODF	Optical distribution frames
ODN	Optical distribution network

OFDM	Orthogonal frequency division multiplexing
OFDMA	Orthogonal frequency division multiple access
OLT	Optical line terminal
OMCC	ONU management communication channel
OMCI	ONU management and control interface
ONU	Optical network unit
OPEX	Operational expenditure
OSI	Open systems interconnection
P2MP	Point to MultiPoint
PCBd	Downstream physical control block
PCS	Physical coding sublayer
PLC	Power line communication
PLOAM	Physical layer OAM
PMA	Physical media attachment
PMD	Physical media dependent
PON	Passive optical network
POTS	Plain old telephone service
QoE	Quality of experience
QoS	Quality of service
RADSL	Rate-adaptive DSL
RNC	Radio network controller
RRU	Remote radio unit
RS	Reconciliation sublayer
RTT	Round trip time
SAE GW	System architecture evolution gateway
SDSL	Symmetric DSL
SERDES	Serializer/deserializer
SOA	Semiconductor optical amplifier
TCONT	Transmission container
TDM	Time division multiplexing
UMTS	Universal mobile telecommunications system
UNI	User network interface
UWB	Ultra-WideBand
VHDSL	Very high-bit-rate DSL
VoIP	Voice over internet protocol
WDM	Wavelength division multiplexing
WLAN	Wireless local area network
WMAN	Wireless metropolitan area network
WWAN	Wireless wide area network

Chapter 1
Overview of Broadband Access Technologies

Over the past decade of unprecedented advances in information and communications technology (ICT), a variety of bandwidth-demanding applications, including Internet access, e-mail, e-commerce, voice over internet protocol (VoIP), video conferencing, Internet Protocol Television (IPTV), and online gaming, have emerged and been rapidly deployed in the network. As the Internet traffic grows, it is becoming urgent to efficiently manage, move, and store increasing amount of mission-critical information, thus accelerating the demand for data storage systems. Consequently, the traffic in both public and private communication networks has experienced dramatic growth. As reported by Cisco's visual networking index, the Internet traffic in 2011 has reached around 28 k petabytes per month while it was less than 200 petabytes per month in 2001 [1]. According to the sixth annual Cisco(R) Visual Networking Index (VNI) Forecast (2011–2016) [2], global IP traffic has increased eightfold over the past 5 years, and will increase 4-fold over the next 5 years. In 2016, global IP traffic will reach 1.3 zettabytes per year or 109.5 exabytes per month. Overall, IP traffic will grow at a compound annual growth rate (CAGR) of 29% from 2011 to 2016. Therefore, to meet the challenge caused by the increased network traffic, telecommunication service providers and enterprises are driven to enhance their networks in providing enough bandwidth for new arising services.

Access networks are the last mile in the Internet access, and therefore should be upgraded to meet the demand of increasing traffic growth as well. Broadband access network operators are currently trying to find faster, easier, and more cost-efficient ways to increase the network bandwidth. Broadband was defined as a "transmission capacity that is faster than primary rate Integrated Services Digital Network (ISDN) at 1.5 or 2.0 Megabits per second (Mbits)" in ITU specification I.113 [38]. Basic rate ISDN, i.e., ISDN-BRI, contains two bearer channels, each of which provides 64 kbit/s data rate. The two channels can be either used separately for voice or data calls or bonded together to provide a 128 kbit/s service. Therefore, the data rate provisioned by broadband access should be higher than 128 kbit/s.

N. Ansari and J. Zhang, *Media Access Control and Resource Allocation: For Next Generation Passive Optical Networks*, SpringerBriefs in Applied Sciences and Technology, DOI 10.1007/978-1-4614-3939-4_1, © The Authors 2013

However, as the broadband access technologies keep evolving, the definition of broadband access will be changed as well. Currently, the broadband access can be offered over the following communication media: digital subscriber line, hybrid fiber coaxial cable, broadband over powerline, wireless, satellite, and optical fiber. The following describes main characteristics of these access technologies and analyzes their capabilities in catering future Internet services.

1.1 Broadband Access Technologies

1.1.1 Digital Subscriber Line

Digital subscriber line (DSL), as standardized in ITU-T G.922 [122], is provisioned through copper pairs, which were originally installed to deliver fixed-line telephone services. Figure 1.1 shows the general DSL architecture. A DSL modem at each subscriber's home connects to a DSL access multiplexer (DSLAM) located at the central office via a dedicated copper pair. In order to be separate from plain old telephone service (POTS), DSL uses the frequency band from 4,000 Hz to as high as 4 MHz. At the subscriber's site, a DSL filter in each outlet removes the high frequency interference, and thus enables simultaneous transmission of voice and data.

There are currently six different types of DSL: Asymmetric DSL (ADSL), Symmetric DSL (SDSL), ISDN DSL (IDSL), High-bit-rate DSL (HDSL), Very high-bit-rate DSL (VDSL), and Rate-Adaptive DSL (RADSL). Each one presents different technical ranges, limitations, and provisioning data rates.

The downstream data rates provisioned by these DSL services range from 256 kbit/s to 40 Mbit/s. ADSL2+, as standardized in ITU G.992.5 [41], can support as high as 24.0 Mbit/s downstream data rate, while IDSL offers a limited data rate of 128 kbps. The upstream data rate is different for different DSL technologies as well. In ADSL, the data rate in the upstream direction (the direction from the subscriber

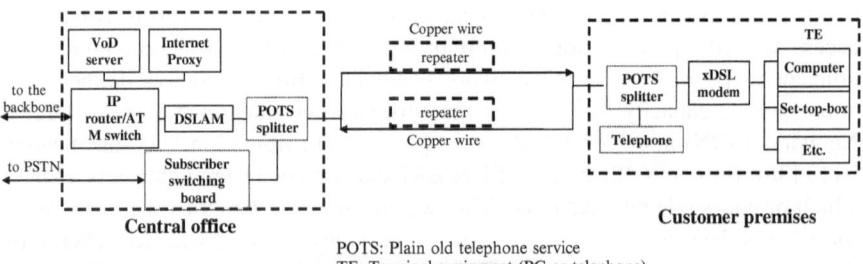

POTS: Plain old telephone service
TE: Terminal equipment (PC or telephone)
DSLAM: DSL access multiplexer

Fig. 1.1 DSL architecture

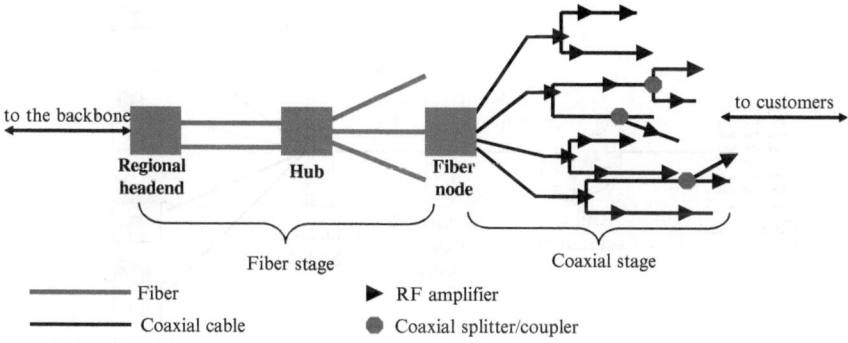

Fig. 1.2 HFC network

to the service provider) is lower than that of the downstream direction (the direction from the service provider to the subscriber). Symmetric DSL (SDSL) services offer equal downstream and upstream data rates.

1.1.2 Hybrid Fiber Coaxial Cable

Hybrid fiber coaxial cable (HFC), which combines optical fiber and coaxial cable, has been commonly employed by cable television operators since the early 1990s [68]. It is standardized in ITU-T J.112/122 [39, 40]. Figure 1.2 illustrates an HFC network. The fiber optic network extends from the cable operator's master headend, sometimes to regional headends, and out to a neighborhood's hubsite, and finally to a fiber optic node. A master headend is usually facilitated with IP aggregation routers as well as satellite dishes for reception of distant video signals. A regional or area headend/hub receives the video signal from the master headend and adds to it the public, educational, and government (PEG) access cable TV channels as required by local franchising authorities. The various services are encoded, modulated and upconverted onto radio frequency (RF) carriers, and combined onto a single electrical signal. The single electrical signal is further inserted into a broadband optical transmitter, and then distributed to customers through a tree network of coaxial cables, with electrical amplifiers placed as necessarily in the network to maintain signal quality. Hence, these networks are commonly termed hybrid fiber coaxial networks.

1.1.3 Broadband Over Powerline

Broadband over powerline (BPL) allows relatively high-speed digital data transmission over the public electric power distribution wiring to provide access to the Internet [107]. Figure 1.3 illustrates the BPL architecture.

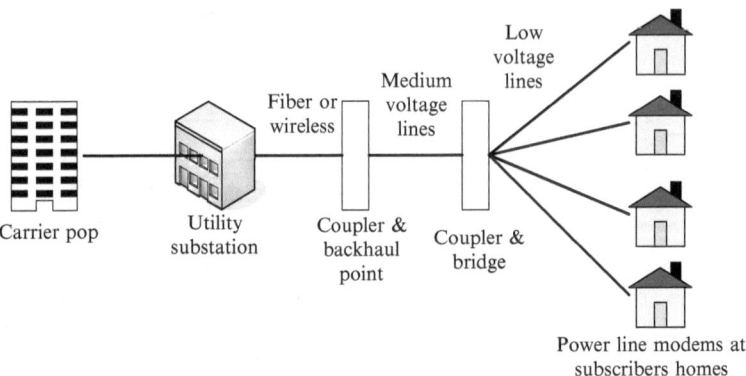

Fig. 1.3 BPL architecture

In BPL systems, the downstream data are first transmitted over traditional fiber-optic lines downstream, and then transmitted onto medium-voltage power lines. Once the traffic is carried in the medium-voltage lines, its transmission range is limited owing to the signal degradation in the power lines. To overcome this problem, repeaters are installed in the lines to amplify the signal strength.

PLC (Power line communication) modems located at subscribers' homes transmit in medium and high frequencies (1.6–80 MHz electric carrier). Typically, the modem supports asymmetric data rates ranging from 256 kbit/s to 2.7 Mbit/s. A repeater situated in the meter room can provision speed of up to 45 Mbit/s and can be connected to 256 PLC modems. In the medium voltage stations, the speed from the head ends to the Internet can be provisioned at up to 135 Mbit/s. To connect to the Internet, utilities can use an optical fiber backbone or a wireless link.

BPL may offer benefits over regular cable modems or digital subscriber line (DSL) connections. BPL allows the Internet access in the remote areas with relatively little equipment investment since the extensive infrastructure is already available. However, interference from these services may limit the data rate of BPL systems, and variations in the characteristics of the physical channel of the electricity network and the lack of standards limit the deployment of the BPL services.

1.1.4 Wireless Broadband Access

Wireless broadband access networks provide wireless service comparable to that of wireline networks for fixed and mobile users [35]. Generally, according to their coverage areas, as shown in Fig. 1.4, wireless data networks can be classified into three categories: Wireless Local Area Networks (WLANs), Wireless Metropolitan Area Networks (WMANs), and Wireless Wide Area Networks (WWANs). WLANs

Fig. 1.4 Broadband wireless networks

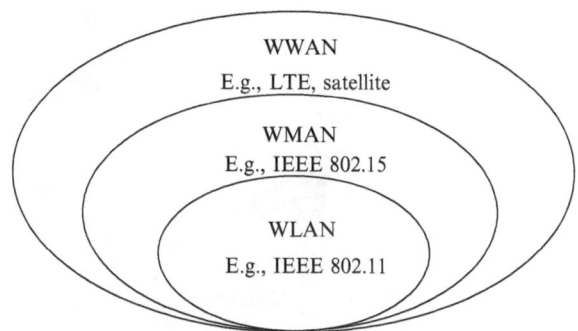

provide wireless access in areas with cell radius up to hundreds of meters, and are used mostly in home and office environments. WMAN covers wider areas, generally as large as an entire city, while WWAN may cover multiple cities.

The most notable WLAN standard is the IEEE 802.11 family. It can serve an area with radius of 50–100 m. Although WiFi initially provided an aggregate throughput of 11 Mbps, the current standard, IEEE 802.11ac [34], enables multi-station WLAN throughput of at least 1 Gbit/s and a maximum single link throughput of at least 500 Mbit/s.

The IEEE 802.16 standard was specified in 2001 to provide broadband wireless services to offer a metropolitan area network (MAN) with a radius of about 50 km (30 miles). The original Broadband Wireless Access (BWA) can provision fixed Line of Sight (LOS) Subscriber Stations (SSs) from a Base Station (BS). IEEE 802.16-2005 also supports non-LOS (NLOS) SSs and mobile subscribers (MSs). Advanced Air Interface in the current standard IEEE 802.16m provisions data rates of 100 Mbit/s for mobile users and 1 Gbit/s for fixed subscribers.

The main WWAN technologies include satellite communications and mobile telecommunication cellular networks such as LTE, UMTS, CDMA2000, and GSM. For cellular networks, the current 4G LTE Advanced network can provide peak data rate as high as 1 Gb/s. For satellite communications, satellite systems have a wide range of different features and technical limitations, which can greatly affect their usefulness and performances in specific applications. Low Earth Orbit (LEO) satellites can provide satellite services world-wide. However, the data rate offered by LEO satellites is limited. LEO satellites are usually used for voice traffic delivery. Geostationary Earth Orbit (GEO) satellites offer higher data speeds, and are used for television transmission and high speed data transmission. However, their signals cannot reach some polar regions of the world. As shown in Fig. 1.5, the distance between a LEO satellite and the earth surface is less than 2,000 km, and that between a GEO satellite and the earth surface is 35,786 km. Owing to the long distance between a GEO satellites and the earth, it takes a radio signal around 250 ms to travel to the satellite and back to the ground. The high latency experienced by the geostationary satellite communications may inhibit the extensive deployment of GEO satellite communications although it can offer downloading data rate of up to 1 Gbit/s, and uploading data rate of up to 10 Mbit/s.

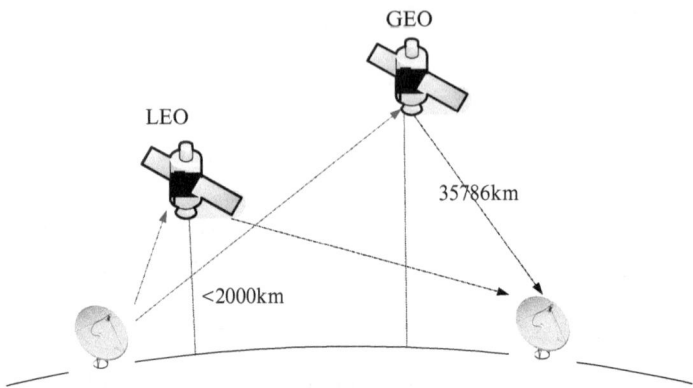

Fig. 1.5 Satellite networks

1.1.5   Optical Fiber

Fiber-to-the-*x*, including fiber-to-the-home (FTTH), fiber-to-the-premise (FTTP), fiber-to-the-building (FTTB), fiber-to-the-node (FTTN), and fiber-to-the-curb or cabinet (FTTC), is the generic term for any broadband network architecture using optical fiber to replace all or part of the usual metal local loop used in the last mile communications [43].

- FTTN (fiber-to-the-node): Fiber is terminated in a street cabinet which can be up to several kilometers away from the customer premises. The final connection between the street cabinet and the subscribers' home is through copper. FTTN is often seen as an interim step towards full fiber-to-the-home (FTTH) and is currently used by telecoms service providers such as AT&T, Deutsche Telekom, Telekom Austria, Belgacom, and Swisscom to deliver advanced triple-play services.
- FTTC (fiber-to-the-curb or fiber to the cabinet): Similar to FTTN, the fiber is terminated at a street cabinet in FTTC, but with a closer distance between the street cabinet or pole and the subscriber's premises. The distance is typically within the range for high bandwidth copper technologies such as wired Ethernet, powerline communication, and wireless WiFi technology.
- FTTB (fiber-to-the-building or fiber-to-the-basement): With the FTTB technology, the optical fiber reaches the boundary of the building, such as the multi-dwelling unit. Similar to FTTN and FTTC, the final connection to the subscriber's home is through other wireline or wireless technologies.
- FTTH (fiber-to-the-home): The optical fiber reaches all the way to the subscriber's premise by FTTH. Usually, the fiber is terminated at a box on the outside wall of a home.
- FTTP (fiber-to-the-premises): FTTP may refer to both FTTH and FTTB, or where the fiber network includes both homes and small businesses.

- FTTD (fiber-to-the-desk): Fiber connection is installed in a computer terminal or fiber media converter near the user's desk.

The currently deployed optical access networks provide 1 Gb/s data rate in both upstream and downstream. IEEE and ITU have respectively specified 10 Gb/s optical access networks. ITU is currently working on the future optical access networks with as high as 40G bit/s data rate provisioning. Owing to the high bandwidth provisioning, the penetration rate of optical fiber keeps increasing year by year globally. According to IDATE [42], fiber-based broadband subscribers increased to nearly 67 millions worldwide within a year by the middle of 2011. The number of buildings and homes laid with fiber networks increased over 47% to almost 179 millions over the same period. As service providers roll out new networks in a bid to cope with surging Internet traffic, the future will see a significant growth of FTTx subscribers. Worldwide subscriber numbers are expected to grow to 198.27 millions by 2015.

Owing to the fact that optical fiber provides higher capacity than other broadband access solutions, optical fiber will be extensively employed in the access network to accommodate future bandwidth demanding applications, and next-generation broadband access will be based on fiber-rich infrastructure and technologies. This book mainly focuses on the optical access network, especially the media access control and resource allocation problems along with their solutions.

1.2 Optical Access Networks

Optical access networks feed the metro and core networks by gathering data from the end subscribers. They connect computers at subscribers' homes with optical line terminals at central offices. Generally, there are three main types of optical access networking technologies: point-to-point fiber, active Ethernet network, and passive optical network (PON) [33]. These three types of optical access networks are different in the devices installed between the central office and the subscriber's home. Figure 1.6 depicts these three networks.

1.2.1 Point-to-Point Fiber

An access network is referred to as point-to-point (PtP) fiber when a direct fiber connection exists between the central office and the optical network termination (ONT) located at the subscriber's home.

As shown in Fig. 1.6a, the outside plant of a PtP fiber architecture contains at least a dedicated fiber terminated at each user's home. Since the fiber is dedicated to each user, the optical power experiences small loss, and the power budget allows

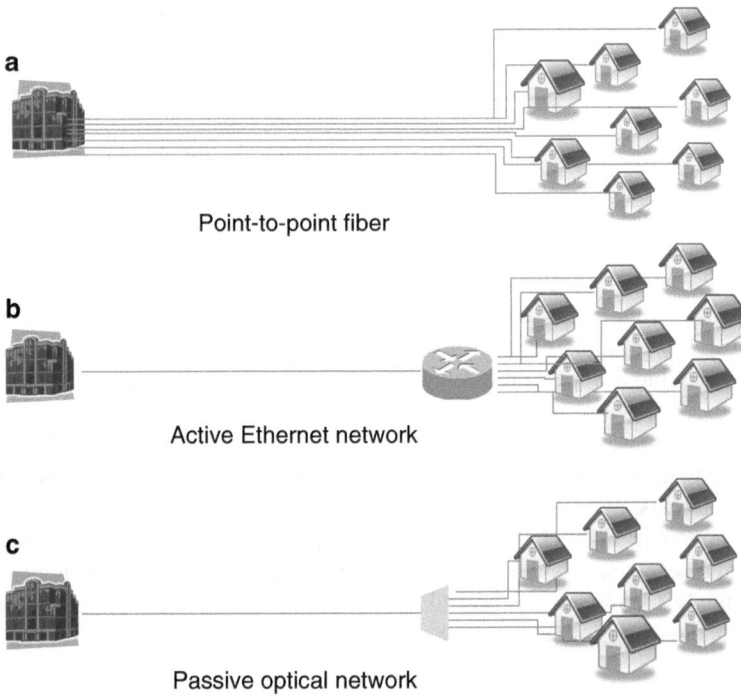

Fig. 1.6 Optical access networks

the distance between the central office and user's home to be as long as 10 km. The link budget is independent of the number of connected users to the OLT because each user has a dedicated port. The simple network architecture and the low power loss eliminate the need of expensive optical components in the network.

The key design parameters of PtP fiber include high-density cables, fiber termination and handling, and port density. The port and fiber densities linearly depend on the size of the central office. Typically, operators employ a combination of optical distribution frames (ODF) to (1) terminate all fibers, and (2) employ patch cords to connect users to an available port at the OLT. Consequently, incremental growth is realized, and low capital and operational expenses are maintained at low take rates.

PtP fiber is future-proof optical access architecture and technology that inherently allows open access and unprecedented bandwidth growth. It is a common way of delivering triple (and quad) play (voice, video, data, and mobile) services. This approach has become increasingly popular in recent years with telecommunications service providers in both North America and Europe. For example, Google has adopted this approach to deliver multiple services over open-access networks in the United States.

1.2.2 Active Ethernet Network

Active Ethernet network uses electrically powered equipment, such as switches and routers, to distribute a signal. This provides TV operators and telecommunications service providers with complete control of their infrastructure, enabling them to guarantee quality of service (QoS) for subscribers.

Similar to PtP fiber, active Ethernet can dedicate each customer with even a 1 Gb/s Ethernet connection to the subscriber. This provides enough bandwidth to deliver all current triple-play services and more. Optical fiber can be easily installed, and the power budget can be easily allocated. Also, as active Ethernet is point-to-point, there is no limitation on the ratio between the distance from the central office to the optical splitter and the distance from the splitter to home. In addition, since the bandwidth in active Ethernet is dedicated, service providers have full control over bandwidth distribution and can promise and deliver different QoS levels.

As depicted in Fig. 1.6b, active Ethernet network can dramatically reduce the number of fibers terminated in the central office while enjoying many advantages of PtP fiber. It has the potential to have the fewest fiber terminations in the central office. However, it requires high investment in the outside plant. To aggregate fibers delivered directly to subscribers, active Ethernet requires switches to be installed in secured cabinets between the central office and subscriber homes.

1.2.3 Passive Optical Network

Similar to active Ethernet network, passive optical network (PON) has a point to multipoint network architecture as depicted in Fig. 1.6 [18]. However, it does not require outside plant electronics. Rather than placing an Ethernet switch at the outside plant, PON uses a passive optical splitter instead. In the downstream, the splitter divides the light sending from the central office and then broadcasts it to all ONUs. In the upstream, the splitter combines the light coming from ONUs, and transmits it over the fiber connected to the OLT. Since there is no optical repeaters or other active devices in the network, the network is referred to as passive optical network.

The passive nature of the splitter in the outside plant allows zero power consumption and maintenance of the outside plant, which is a great advantage over active Ethernet network. In addition, owing to the low price of the splitter, the cost of the outside plant of PON is significantly reduced as compared to active Ethernet network.

In general, PON exhibits four other major advantages. First, PON yields a small fiber deployment cost in the local exchange and local loop. Second, PON provisions higher bandwidth due to the deep fiber penetration. Third, as a point-to-multipoint network, PON can easily facilitate downstream video broadcasting. Fourth, PON eliminates the necessity of installing multiplexers and demultiplexers in the splitting locations, and thus lowers the operational expenditure.

In this book, we will discuss solutions to media access control and resource allocation problems in various PON architectures which are the promising next-generation broadband access network architectures.

1.3   Summary

This chapter overviews the main broadband access technologies including digital subscriber line, hybrid fiber coaxial cable, broadband over powerline, wireless broadband access, and optical fibers. Digital subscriber line provisions the downstream data rate ranging from 256 kbit/s to 40 Mbit/s. ADSL2+ can support as high as 24.0 Mbit/s downstream data rate. Hybrid fiber coaxial cable combines optical fiber and coaxial cable. Broadband over powerline (BPL) facilitates the high-speed digital data transmission over the public electric power distribution wiring. It is an attractive solution to provide Internet access in remote areas since the extensive infrastructure is already available. Empowered by recent advances in wireless access technologies, wireless broadband access networks provide wireless service comparable to that of wireline networks for fixed and mobile users. Optical fiber provides the highest capacity among all broadband access solutions. Owing to the high bandwidth provisioning, optical access networks will be extensively employed in the future to accommodate the growing traffic demands. This chapter also discusses three main optical access technologies: point-to-point fiber, active Ethernet network, and passive optical network. These three types of optical access networks are different in the devices installed between the central office and the subscriber's home.

Chapter 2
PON Architectures

Passive Optical Network (PON) [112] is a set of technologies standardized by ITU-T and IEEE, although it is originally created by the Full Service Access Network (FSAN) working group. PON is a converged infrastructure that can carry multiple services such as plain old telephony service (POTS), voice over IP (VoIP), data, video, and/or telemetry, in that all of these services are converted and encapsulated in a single packet type for transmission over the PON fiber. PON consists of three main parts [63].

- Optical Line Terminal (OLT): The OLT is located at the service provider's central office. It provides the interface between PON and the backbone network.
- Optical Network Unit (ONU): The ONU is located near end users. It provides the service interface to end users.
- Optical Distribution Network (ODN): The ODN in PON connects the OLT at the central office and ONUs near user homes by using optical fibers and splitters. The ODN usually forms a tree structure with the OLT as the root of the tree and ONUs as leaves of the tree.

There are three main types of PONs depending on the data multiplexing scheme. The currently deployed PON technology is time division multiplexing (TDM) PON, where traffic from/to multiple ONUs are TDM multiplexed onto the upstream/downstream wavelength. Wavelength division multiplexing (WDM) PON and orthogonal frequency division multiplexing (OFDM) PON constitute another two types of PON technologies [78, 111]. WDM PON uses multiple wavelengths to provision bandwidth to ONUs, while OFDM PON employs a number of orthogonal subcarriers to transmit traffic from/to ONUs. With the WDM or OFDM technology, these PONs are potentially able to provide higher than 40 Gb/s data rate and even Tera bps data rate.

Figure 2.1 shows the evolution of PONs. TDM is the multiplexing scheme for all PONs which have been standardized until now. Numerous research articles in tackling challenges in WDM PON have been published since WDM PON was first proposed in 1986 [86]. However, it has not been included in the standard yet mainly

N. Ansari and J. Zhang, *Media Access Control and Resource Allocation: For Next Generation Passive Optical Networks*, SpringerBriefs in Applied Sciences and Technology, DOI 10.1007/978-1-4614-3939-4_2, © The Authors 2013

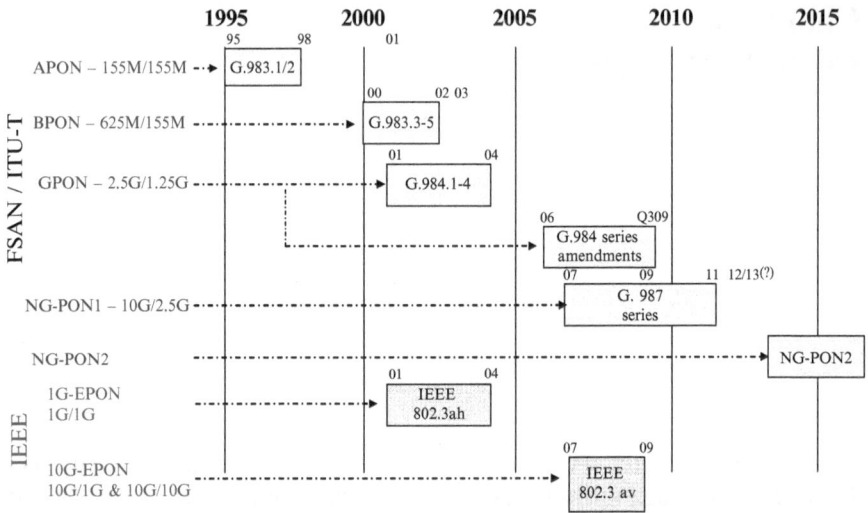

Fig. 2.1 Evolution of PON

due to the high system cost. OFDM PON has received intensive research attention in recent years owing to its high bandwidth provisioning. Both WDM PON and OFDM PON are considered as future PON technologies.

2.1 TDM PON

The currently deployed PON systems are TDM PON systems, which include ATM PON (APON), Broadband PON (BPON), Ethernet PON (EPON), Gigabit PON (GPON), 10G EPON, and Next-generation PON (NG-PON) to provision different data rates [18, 27, 55].

APON/BPON, GPON, and NG-PON architectures were standardized by the Full Service Access Network (FSAN), which is an affiliation of network operators and telecom vendors. Since most telecommunications operators have heavily invested in providing legacy TDM services, these PON architectures are optimized for TDM traffic and rely on framing structures with a very strict timing and synchronization requirements.

EPON and 10G-EPON are standardized by the IEEE 802 study group. They focus on preserving the architectural model of Ethernet. No explicit framing structure exists in EPON, and Ethernet frames are transmitted in bursts with a standard inter-frame spacing.

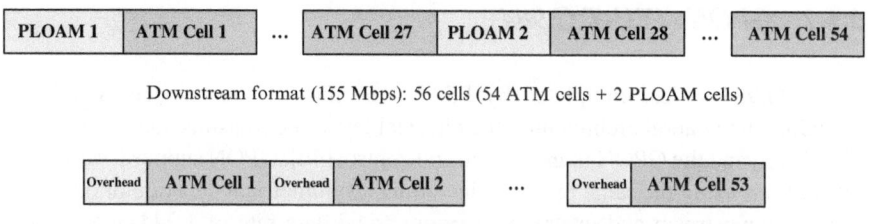

Downstream format (155 Mbps): 56 cells (54 ATM cells + 2 PLOAM cells)

Upstream format (155 Mbps): 53 cells (53-byte ATM cell + 3-byte Overhead)

Fig. 2.2 APON frame format

2.1.1 APON/BPON (ITU-T G.983)

APON, ATM PON, is the initial PON specifications defined by the FSAN committee. APON uses ATM as their signaling protocol in layer 2. In APON, downstream transmission is a continuous ATM stream at a bit rate of 155.52 Mb/s or 622.08 Mb/s. Upstream transmission is in the form of bursts of ATM cells.

Figure 2.2 shows APON frame formats. The upstream channel is divided into 53 slots of 56 bytes at 155.520 Mbps, while the downstream cell stream is divided into frames of 56 cells at 155.520 Mbps. Physical Layer Operation, Administration and Maintenance (PLOAM) cells are inserted at the beginning and in the middle of a downstream frame. Each PLOAM cell contains 27 grant fields and a 12-byte message field. Grant fields are used to control the upstream data transmission, and message fields are used to control the operation of the ONUs.

In the upstream frame, a 3-byte overhead header is transmitted before a 53-byte ATM cell in each slot. The 3-byte overhead contains a minimum of four bits of guard time, a preamble, and a delimiter field. The guard time is to ensure a sufficient distance between two continuous cells to prevent collisions. The preamble field is used to extract the phase of the incoming ATM cell and acquire bit synchronization. The delimiter field is a unique bit pattern indicating the start of an incoming cell. PLOAM cells may be transmitted instead of ATM cells in the upstream frame to convey the physical layer information of an ONU to the OLT.

Broadband PON (BPON), as defined in ITU-T G.983 series, is a further improvement of the APON system. With the objective of achieving early and cost-effective deployment of broadband optical access systems, BPON offers numerous broadband services including ATM, Ethernet access, and video distribution. BPON employs wavelength division multiplexing (WDM) for downstream transmission, with as many as 16 wavelengths with 200 GHz spacing or 32 wavelengths with 100 GHz spacing between channels. BPON also provides enhanced security through the churning technique in which the encryption key is changed at least once a second between the Optical Line Terminal (OLT) at the headend and the Optical Network Terminal (ONT) at the customer premises.

2.1.2   GPON (ITU-T G.984)

ITU-T G.984 series, completed by FSAN, specifies various aspects of GPON, including the general architecture, the physical layer, the transmission convergence (TC) layer, and the GPON management and control [46]. GPON supports various bit rate options using the same protocol, including a symmetrical data rate of 622 Mb/s in both downstream and upstream, a symmetrical data rate of 1.244 Gb/s in both streams, as well as a data rate of 2.488 Gb/s in downstream and a data rate of 1.244 Gb/s in upstream. 2.488 Gb/s of downstream bandwidth and 1.244 Gb/s of upstream bandwidth are the data rates supported by typical GPON systems.

GPON defines the GPON encapsulation method (GEM) to achieve efficient packaging of user traffic, with frame segmentation to better provide quality of service (QoS) for delay-sensitive traffic such as voice and video applications. It accommodates three layer-2 networks: ATM for voice, Ethernet for data, and proprietary encapsulation for video, thus enabling GPON with full-service support capability, including voice, time division multiplexing (TDM), Ethernet, ATM, leased lines, and wireless extension. GPON also supports radio frequency (RF) video transmission in the waveband from 1,550 to 1,560 nm.

GPON directly reflects the requirements of network operators because the GPON standardization is driven by operators through FSAN. Similar to APON/BPON, GPON provides high product interoperability by standardizing a management interface, referred to as the optical network unit management and control interface (OMCI), between OLTs and ONUs/ONTs. It provides strong Operation Administration, Maintenance, and Provisioning (OAM&P) capabilities offering end-to-end service management.

2.1.3   NG-PON (ITU-T G.987)

Having completed the mission on GPON, ITU-T/FSAN has since been investigating next-generation PON (NG-PON) with higher bandwidth provisioning [135]. The evolution of NG-PON is divided into two phases: NG-PON1 and NG-PON2. NG-PON1 focuses on PON technologies that are compatible with GPON standards (ITU-T G.984 series) as well as the current optical distribution network (ODN). NG-PON1 is backwardly compatible with existing fiber installations, and tries to facilitate high bandwidth provision, large split ratio, and extended network reach. The objective of NG-PON2 is to provision an independent PON system, without being constrained by the GPON standards and the currently deployed outside plant.

As being standardized in ITU-T G.987, NG-PON1 specifies both asymmetric and symmetric 10G-PONs [50]. Figure 2.3 illustrates the NG-PON1 architecture. Asymmetric 10G-PON, also referred to as XG-PON1, provides the downstream data rate of 9.95328 Gbit/s and the upstream data rate of 2.48832 Gbit/s. This architecture upgrades the downstream link capacity to 10 Gb/s. The difficulty of the architecture in provisioning 10 Gb/s is to enable the burst mode time-division multiple access

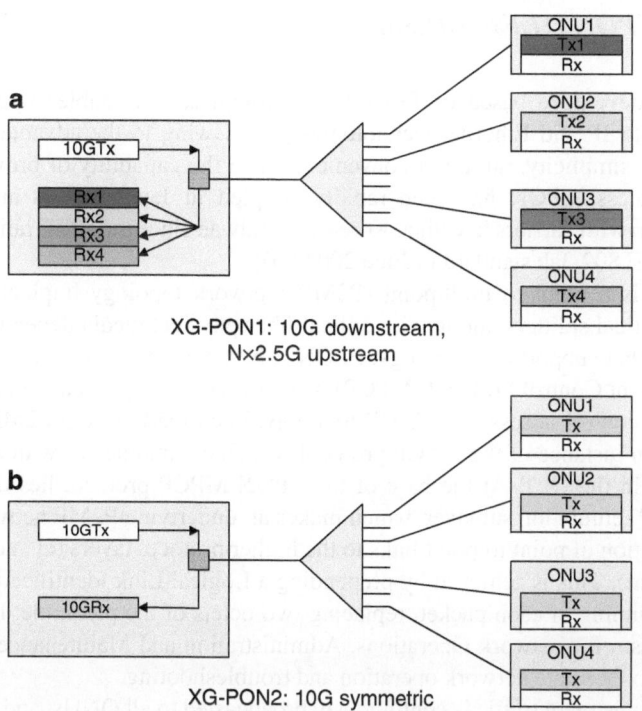

Fig. 2.3 XG-PON architecture

(TDMA) operated at 10 Gb/s. Owing to the limitation of available components and design practices, many simple circuit techniques become impractical when the rate goes beyond 5 Gb/s. Overcoming this limit requires specialized hardware and is thus costly.

To minimize the incurred additional investment, an architecture was proposed to upgrade only the downstream to 10 Gb/s, but to use one or more 2.5 Gb/s wavelengths in the upstream as shown in Fig. 2.3a. This architecture can still be considered as a TDM system both in the downstream and upstream. The downstream transmission can be modeled as 32 ONUs sharing a 10 Gb/s link. Depending on the number of available upstream wavelengths, the ONUs in the upstream scenario are divided into a different number of groups operating at 2.5 Gb/s. If two wavelengths are adopted in the upstream, the ONUs in the upstream scenario are divided into two virtual groups, each of which has 16 ONUs sharing a 2.5 Gb/s upstream link. If one wavelength is adopted, it can be abstracted as 32 ONUs sharing a 2.5 Gb/s upstream link from the MAC layer's perspective.

Symmetric 10G-PON, referred to as XG-PON2, achieves 10 Gbit/s in both upstream and downstream. However, XG-PON2 requires cost-inefficient burst-mode transmitters at ONU sides to deliver the upstream transmission speed. When devices capable of a 10 Gb/s burst mode become commercially available, the architecture with both the downstream and upstream transmission being upgraded to 10 Gb/s can be realized (see Fig. 2.3b).

2.1.4   EPON (IEEE 802.3ah)

EPON is developed based on Ethernet technologies, and enables seamless integration with IP and Ethernet technologies [58]. Owing to the advantages of fine scalability, simplicity, multicast convenience, and the capability of providing full-service access, EPON has been rapidly adopted in Japan and is also gaining momentum with carriers in China, Korea, and Taiwan since the IEEE ratified EPON as the IEEE 802.3ah standard in June 2004 [10].

EPON is a point to multipoint (P2MP) network topology implemented with passive optical splitters, along with optical fiber physical media dependent (PMD) sublayers that support this topology. EPON is based upon a mechanism referred to as MultiPoint Control Protocol (MPCP), which uses messages, state machines, and timers, to control access to a P2MP topology. Each ONU in the P2MP topology contains an instance of the MPCP protocol, which communicates with an instance of MPCP in the OLT. At the base of the EPON MPCP protocol lies the point to point (P2P) emulation sublayer, which makes an underlying P2MP network appear as a collection of point to point links to the higher protocol layers (at and above the MAC client). This is achieved by prepending a Logical Link Identification (LLID) at the beginning of each packet, replacing two octets of the preamble. In addition, a mechanism for network Operations, Administration and Maintenance (OAM) is included to facilitate network operation and troubleshooting.

The downstream traffic is continuously broadcasted to all ONUs, and each ONU selects the packets destined to it and discards packets addressed to other ONUs. In the upstream, each ONU transmits during the time slots that are allocated by the OLT. Upstream signals are combined by using a multiple access protocol, usually time division multiple access (TDMA). The OLT "ranges" the ONUs in order to provide time slot assignments for upstream communications. Owing to their burst transmission nature, burst-mode transceivers are required to fulfill the upstream transmission from an ONU to the OLT.

As compared to GPON, the burst sizes and physical layer overhead are large in EPON. As a result, ONUs do not need any protocol and circuitry to adjust the laser power. Also, the laser-on and laser-off times are capped at 512 ns, a significantly higher bound than that of GPON. The relaxed physical overhead values are just a few of many cost-cutting steps taken by EPON. Another key cost-cutting step of EPON is the preservation of the Ethernet framing format, which carries variable-length packets without fragmentation.

2.1.5   10G EPON (IEEE 802.3av)

Motivated to meeting the emerging high bandwidth demands, the IEEE 802.3av 10G-EPON task force was charged to increase the downstream bandwidth to 10 Gb/s, and to support two upstream data rates: 10 and 1 Gb/s [104, 131]. 10G-EPON supports both symmetric 10 Gb/s downstream and upstream, and asymmetric

10 Gb/s downstream and 1 Gb/s upstream data rates, while 1G-EPON provides only the 1 Gb/s symmetric data rate.

With the focus on the physical layer, the IEEE 802.3av Task Force specifies the reconciliation sublayer (RS), symmetric and asymmetric physical coding sublayers (PCSs), physical media attachments (PMAs), and physical media dependent (PMD) sublayers. Table 2.1 lists several key physical layer features of 10G-EPON [7]. Instead of using the 8B/10B line coding adopted in 1G-EPON, 10G-EPON employs 64B/66B line coding, with which the bit-to-baud overhead is reduced to as small as 3 %. To relax the requirements for optical transceivers, Reed-Solomon code (255, 223) is chosen as the mandatory forward error correction (FEC) code in 10G-EPON to enhance the FEC gain, while Reed-Solomon code (255, 239) is specified as optional for 1G-EPON.

10G-EPON denotes PRX as the power budget for asymmetric-rate PHY of 10 Gb/s downstream and 1 Gb/s upstream, and PR as the power budget for symmetric-rate PHY of 10 Gb/s both upstream and downstream. Each power budget further contains three power budget classes: low power budget (PR(X)10), medium power budget (PR(X)20), and high power budget (PR(X)30). PR(X)10 and PR(X)20 power budget classes are defined in 1G-EPON as well, while PR(X)30, which can support 32-split with a distance of at least 20 km, is an additional one defined in 10G-EPON. For illustrative purposes, we only list the transmitter (Tx) type along with its launch power of 10G-EPON in Table 2.1. As compared to 1G-EPON, advanced transmitters and higher launch power are employed in 10G-EPON to guarantee a sufficient signal-to-noise ratio (SNR) at the receiver side for accurate recovery of data with a rate of 10 Gb/s. Because of the increased launch power, the power consumption of the optical transmitter should be increased accordingly. Also, owing to the mandatory FEC mechanism and increased line rate, the electronic circuit has to enable more functions and process faster than that in 1G-EPON, thus consequently incurring higher power consumption and possibly larger heat dissipation. Therefore, to accommodate 10 Gb/s in the physical layer, the power consumption of the OLT and the ONU may increase significantly.

For the MAC layer and layers above, in order to achieve backward compatibility such that network operators are encouraged to upgrade their services, 10G-EPON keeps the EPON frame format, MAC layer, MAC control layer, and all the layers above almost unchanged from 1G-EPON. This further implies that similar network management system (NMS), PON-layer operations, administrations, and maintenance (OAM) system, and dynamic bandwidth allocation (DBA) used in 1G-EPON can be applied to 10G-EPON as well.

Table 2.1 lists the specifications on data rates of several PON standards.

2.2 WDM PON

WDM PON is a candidate solution for next-generation PON systems in competition with 10G-EPON and NG-PON1 systems [60, 65]. To achieve high bandwidth provisioning, WDM PON supplies each subscriber with a wavelength rather than

Table 2.1 Data rate specifications of various PON standards

	APON/BPON	GPON	XG-PON1	EPON	10G-EPON
Standard	ITU-T G.983	ITU-T G.984	ITU-T G.987	IEEE 802.3ah	IEEE 802.3av
Downstream speeds	622 Mbps	2.488 Gbps	9.9528 Gbps	1.25 Gbps	10.3125 Gbps
Upstream speeds	155 Mbps	1.244 Gbps	2.488 Gbps	1.25 Gbps	1.25 Gbps
			9.9528 Gbps		10.3125 Gbps

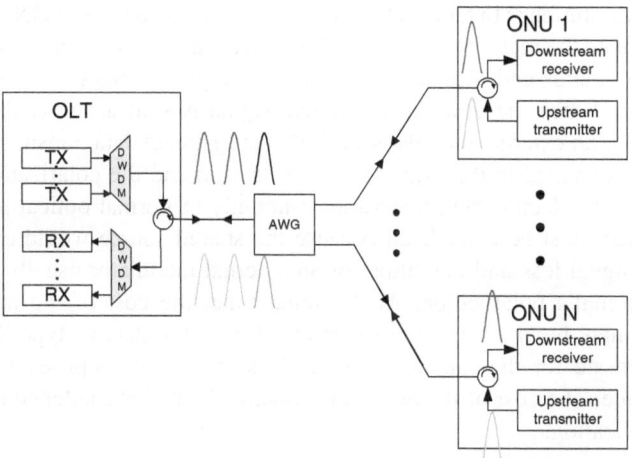

Fig. 2.4 A typical WDM PON architecture

sharing wavelength among 32 or even more subscribers in TDM PON. Figure 2.4 shows a typical WDM PON architecture. As early as 2006, WDM PON has already been deployed in Korea. However, owing to its high costs as compared to EPON and GPON, the deployment of WDM PON in other countries has been stalled.

WDM PON architecture enjoys several advantages over conventional TDM PON systems. First, WDM PON allows each user being dedicated with one or more wavelengths, thus allowing each subscriber to access the full bandwidth accommodated by the wavelengths. Second, WDM PON networks typically provide better security and scalability because each home only receives its own wavelength. Third, the MAC layer control in WDM PON is more simplified as compared to TDM PON because WDM PON provides P2P connections between the OLT and the ONU, and does not require the Point-to-Multipoint (P2MP) media access controllers found in other PON networks. Finally, each wavelength in a WDM PON network is effectively a P2P link, thus allowing each link to run at a different speed and with a different protocol for maximum flexibility and pay-as-you-grow upgrades.

Despite these attractive features, WDM PON is cost inhibitive because of the wavelength specific feature of ONUs. Since each subscriber is dedicated with some wavelengths, the OLT in WDM PON that supports 32 ONUs must transmit on no less than 32 different wavelengths, and each ONU should operate at their own wavelengths. The wavelength-specific feature of ONUs imposes higher requirements on lasers as compared to TDM PONs if the same kind of wavelength fixed lasers for all ONUs is employed. One solution is to use tunable lasers with which each ONU can be tuned to its desired wavelength. However, tunable lasers are costly. Another solution is to equip each subscriber with a wavelength-specific fixed tuned laser. Individual wavelength-specified sources cannot be readily employed in the OLT of WDM PON because they require a number of optical sources with different wavelengths. These cost-prohibitive devices constitute a major hurdle in early design of WDM PON systems.

The third solution is to let the OLT provide all optical sources to ONUs, and each ONU modulate the received unmodulated optical source. Two types of modulators, external modulator and semiconductor optical amplifier (SOA), can be used for this purpose. When the downstream optical signal is split at the ONU, part of it is provided to an external modulator or SOA for upstream data transmission. When the ONU is operated in this way, the power margin and the polarization (i.e., the direction of the electric field that varies randomly in normal optical fiber) of the optical signal must be considered because the shared source would experience a round-trip signal loss and the output of an external modulator usually varies with the input signal's polarization. At the same time, the cost of the modulator at each ONU may be an obstacle to its practical use. A reflective-type SOA, which can compensate for the round-trip signal loss, has been proposed for use as a shared source. The cost of the SOA still remains the major challenge for eventual commercialization.

2.3   OFDM PON

Orthogonal frequency division multiplexing (OFDM) PON[52, 90], as shown in Fig. 2.5, employs OFDM as the modulation scheme and exploits its superior transmission capability to improve the bandwidth provisioning of optical access networks. OFDM uses a large number of closely-spaced orthogonal subcarriers to carry data traffic. Each subcarrier is modulated by a conventional modulation scheme (such as quadrature amplitude modulation or phase-shift keying) at a low symbol rate, thus achieving the sum of the rates provided by all subcarriers compatible to those of conventional single-carrier modulation schemes in the same bandwidth. Since the data rate carried by each subcarrier is low, the duration of each symbol is relatively large. Thus, the inter-symbol interference can be efficiently reduced in a wireless multipath channel. In optical communications, the dispersion including chromatic dispersion and polarization mode dispersion has similar effects as those of multipath. Therefore, employing the OFDM modulation scheme in the optical access network can greatly increase the network provisioning data rate and lengthen the network reach.

OFDM has been successfully applied to ADSL, DVB-T, WLAN and WiMAX, and is a key transmission technology for next generation wireless systems including 3GPP LTE. In OFDM PON, cheaper electronic devices are used instead of costly optical devices, and ASIC-based DSP and AD/DA also reduce equipment costs. OFDM-PON can be combined with WDM to further increase the bandwidth provisioning, and has therefore become a competitive technology for NG-PON2. OFDM PON exhibits the following advantages:

- *Enhanced spectral efficiency*: Orthogonality among subcarriers in OFDM allows spectral overlap of individual subchannels. In addition, OFDM uses a simple constellation mapping algorithm for high-order modulation schemes such as

Fig. 2.5 OFDM PON
architecture

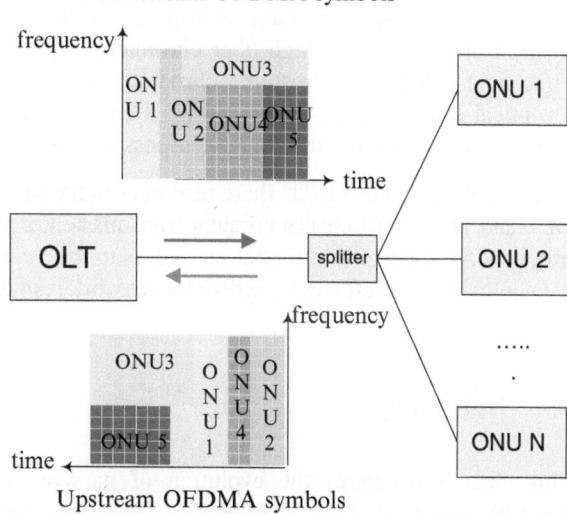

16QAM and 8PSK. Using these techniques, OFDM in PON makes effective use
of spectral resources and improves spectral efficiency.

- *Avoiding costly optical devices and using cheaper electronic devices*: Integrated
  optical devices are very costly, and optical modules of 10G or higher can
  significantly drive up the cost of an access network. OFDM avoids costly
  optical devices and uses cheaper electronic devices. OFDM leverages on the
  integration and low-cost advantages of high-speed digital signal processors and
  high-frequency microwave devices to develop access networks.
- *Dynamic allocation of subcarriers*: Depending on channel environments and
  application scenarios, OFDM can dynamically allocate the number of bits carried
  by each subcarrier, determine the modulation scheme used by each subcarrier,
  and adjust the transmitting power of each subcarrier by using a simple FFT
  algorithm. In OFDM-PON, allocation of each subcarrier is executed in real time
  according to the access distance, subscriber type, and access service.
- *Converged wireline and wireless access*: An optical access network has enormous
  bandwidth potential and good QoS but lacks mobility, and is unable to meet the
  diverse requirements of different terminals. A wireless access network is more
  flexible and provides mobility, but suffers from poor QoS. OFDM is a mature
  technology in wireless communications that has been applied to WiMAX, WiFi
  and UWB (Ultra-wideband). By using OFDM to carry PON signals, wireline
  and wireless access can be converged. In other words, OFDM supports access to
  baseband OFDM, UWB (i.e., MultiBand-OFDM for UWB), WiMAX, WiFi, and
  millimeter-wave OFDM signals. Such versatility has significantly enhanced the
  universality of access networks.
- *Smooth evolution to ultra-long-haul access network*: A simple network structure
  improves the performance of an access network and reduces costs. Converged

optical core, metro, and access network has become a hot research topic, and long reach access networks have been proposed. Long-reach optical access suffers from the problem of high fiber chromatic dispersion. The OFDM modulation scheme can help address the chromatic and polarization-mode dispersion in optical links. Therefore, OFDM-PON can be used to smoothly evolve optical access networks to ultra-longhaul access networks.

In PONs, multiple ONUs share resources in the same feeder fiber connecting the OLT and ONUs. To ensure efficient transmission, a PON system must employ a proper MAC mechanism to arbitrate access to the shared medium in order to avoid data collisions and efficiently utilize the network resources.

2.4   Summary

This chapter discusses the evolution of passive optical networks and briefly describes the major PON technologies. There are three major PON technologies: TDM PON, WDM PON, and OFDM PON. The currently deployed PON networks are TDM PON systems. TDM PON has several flavors including ATM PON (APON), Broadband PON (BPON), Ethernet PON (EPON), Gigabit PON (GPON), 10G EPON, and XG-PON. WDM PON is a candidate solution for next-generation PON systems in competition with TDM PON systems. To achieve high bandwidth provisioning, WDM PON supplies each subscriber with a wavelength rather than sharing wavelengths among 32 or even more subscribers in TDM PON. Orthogonal frequency division multiplexing (OFDM) PON employs OFDM as the modulation scheme and exploits its superior transmission capability to improve the bandwidth provisioning of optical access networks.

Chapter 3
Media Access Control and Resource Allocation in GPON

The protocol stack of GPON systems, as depicted in Fig. 3.1, is specified in ITU-T G.984 series [20, 29, 46]. More specifically,

- ITU-T G.984.2 specifies the GPON physical medium and the physical medium dependents (optics).
- ITU-T G.984.3 defines the transmission convergence layer, which is responsible for constructing the transmission frame, encapsulating the GPON frame by using the GPON encapsulation method (GEM), constructing the Physical Layer Operation, Administration, and Maintenance (PLOAM) channel, and performing the dynamic bandwidth allocation and PON-level QoS control.
- ITU-T G.984.4 defines the ONT management and configuration interface (OMCI). This layer defines (1) the management information base (MIB) for all the functions controlled in the ONU, and (2) the ONU management communication channel (OMCC) that provides all the mechanisms required for the OLT to provide FCAPS (fault, configuration, accounting, performance, security) functionality for the ONU. The OLT management is a somewhat more complex object. It contains by proxy all MIBs of all ONUs supported by that OLT, as well as all the other MIBs that describe the other functions in the OLT.

3.1 Media Access Control in GPON

3.1.1 Downstream Transmission

In the downstream, the OLT centralizes the traffic multiplexing functionality. The OLT multiplexes GEM frames onto the transmission medium by using the GEM Port ID as a key to identify the GEM frames that belong to different downstream logical connections as shown in Fig. 3.2. Each ONU filters the downstream GEM frames based on their GEM Port IDs and processes only the GEM frames that belong to that ONU.

N. Ansari and J. Zhang, *Media Access Control and Resource Allocation: For Next Generation Passive Optical Networks*, SpringerBriefs in Applied Sciences and Technology, DOI 10.1007/978-1-4614-3939-4_3, © The Authors 2013

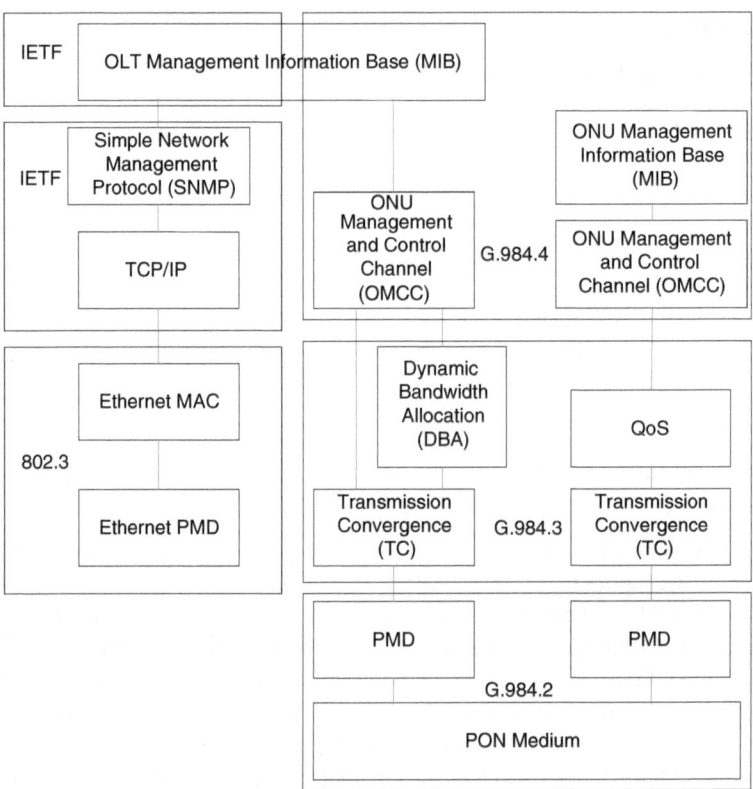

Fig. 3.1 GPON protocol stack [46]

3.1.2   Upstream Transmission

In the upstream direction, the traffic multiplexing functionality is distributed. The OLT grants upstream transmission opportunities, or upstream bandwidth allocations, to the traffic-bearing entities within the subtending ONUs. The traffic-bearing entities of ONUs are recipients of the upstream bandwidth allocations. They are identified by their allocation IDs (Alloc-IDs). The bandwidth allocations to different Alloc-IDs are multiplexed in time as specified by the OLT in the bandwidth maps transmitted downstream. Within each bandwidth allocation, the ONU uses the GEM Port-ID as a multiplexing key to identify GEM frames that belong to different upstream logical connections.

Media access control and resource allocation for upstream traffic are implemented in the GPON transmission convergence (GTC) layer as specified in ITU-T G.984.3. Basically, downstream frames indicate permitted locations for upstream traffic in upstream GTC frames. The media access control concept in the GTC system is illustrated in Fig. 3.3.

Fig. 3.2 The downstream
multiplexing in GPON

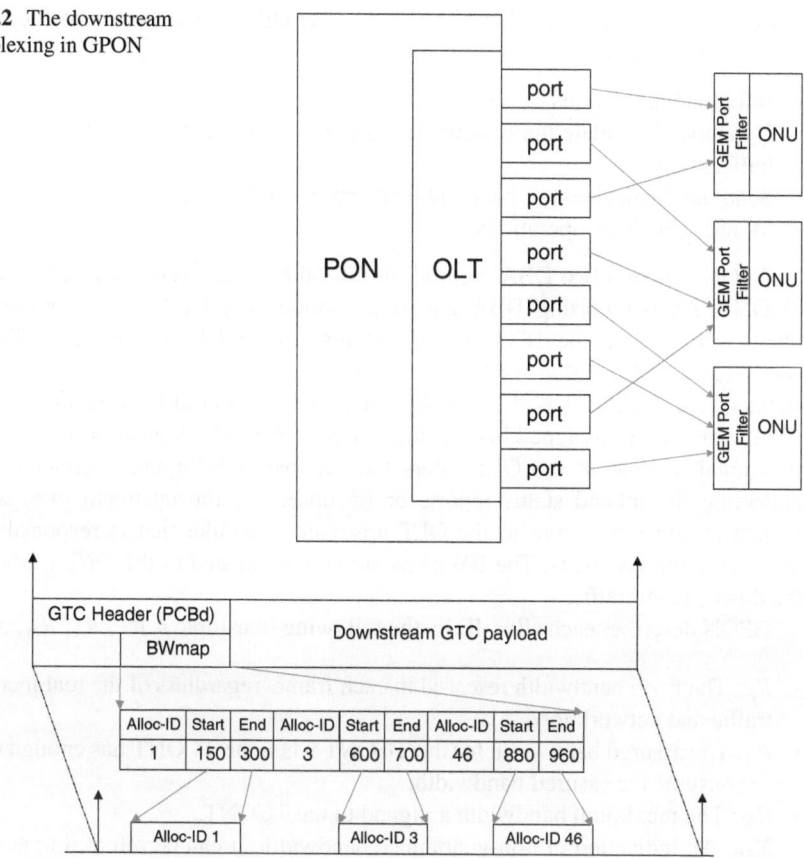

Fig. 3.3 GPON transmission convergence media access control

The OLT sends pointers in the upstream bandwidth map (BWmap) field of the downstream physical control block (PCBd), and these pointers indicate the time at which each ONU may begin and end its upstream transmission. By doing so, only one ONU can access the medium at any time, and there is no contention in normal operation. The pointers are given in units of bytes, allowing the OLT to control the medium at an effective static bandwidth granularity of 64 kbit/s. However, some implementations of the OLT may choose to set the values of the pointers at a larger granularity, and to achieve fine bandwidth control via dynamic scheduling.

3.2 Dynamic Bandwidth Allocation

The basic control unit of bandwidth allocation is a transmission container (TCONT), indexed by allocation ID (alloc-ID). The process in which the OLT dynamically allocates upstream bandwidth to TCONTs, based on their traffic loads, is referred to as dynamic bandwidth allocation (DBA). The DBA process contains four major functional modules.

- Infer the buffer status of each ONU.
- Dynamically update the bandwidth assigned to ONUs based on their real-time buffer status.
- Send out the decision of bandwidth allocations to ONUs.
- Manage the DBA operations.

GPON supports two DBA methods to infer the buffer occupancy status of each TCONT: status-reporting DBA and traffic-monitoring DBA. In status-reporting DBA, ONUs report the TCONT buffer status to the OLT by directly sending the status report information. In traffic monitoring DBA, the OLT infers the buffer status of the TCONT based on the historical information of bandwidth utilization and the amount of assigned bandwidth. For each Alloc-ID logical buffer, the DBA functional module of the OLT infers the occupancy information either through collecting the inband status reports or by observing the upstream idle pattern. It then provides the input to the OLT upstream scheduler that is responsible for generating the BWmaps. The BWmaps are communicated to the ONUs inband in the downstream traffic.

GPON describes each alloc-ID by the following four-tuple $< R_F, R_A, R_M, X_{AB} >$.

- R_F: The fixed bandwidth reserved in each frame, regardless of the real incoming traffic and network load.
- R_A: The assured bandwidth for the TCONT when the TCONT has enough traffic to consume the assured bandwidth.
- R_M: The maximum bandwidth assigned to the TCONT.
- X_{AB}: An indication of non-guaranteed bandwidth. It can be referred to as either non-assured bandwidth or best-effort bandwidth.

The DBA in GPON follows a strict priority hierarchy: fixed bandwidth, assured bandwidth, non-assured bandwidth, and best-effort bandwidth (see Fig. 3.4). First, the OLT assigns the upstream bandwidth to the fixed bandwidth requirement of each TCONT. Then, the OLT allocates bandwidth to the assured bandwidth components of each TCONT, as long as the TCONT has enough traffic to consume the assured one. Third, the OLT satisfies the requirement of non-assured bandwidth. Last, the OLT allocates the remaining bandwidth to the best-effort bandwidth component. The assignment of the non-assured bandwidth and best-effort bandwidth adopts the following rules:

- For non-assured bandwidth, the OLT assigns bandwidth to the TCONT in proportion to the sum of the fixed bandwidth and assured bandwidth of that TCONT.
- For the best-effort bandwidth, the OLT assigns bandwidth to the TCONT in proportion to the value of the maximum bandwidth minus the guaranteed bandwidth of that TCONT (the sum of fixed bandwidth and assured bandwidth).

Fig. 3.4 Bandwidth
allocation for different traffic
loads

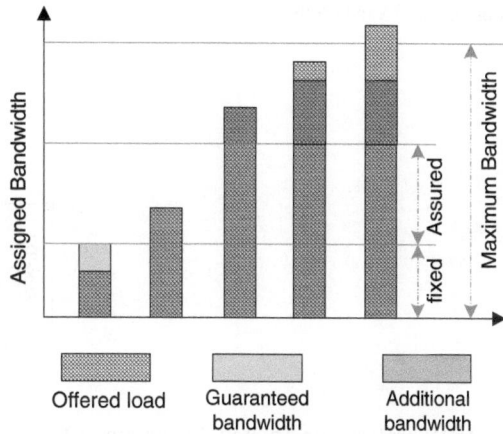

3.3 **Traffic Mapping**

GPON further defines five TCONT types with different combinations of traffic
descriptors:

- TCONT type 1: Fixed bandwidth component only, suitable for constant bit-rate
  (CBR) traffic that is delay and jitter sensitive.
- TCONT type 2: Assured bandwidth component only, suitable for traffic that does
  not have strict delay and jitter bounds.
- TCONT type 3: Assured bandwidth component plus non-assured bandwidth
  component, suitable for variable-rate, bursty traffic that requires an average rate
  guarantee.
- TCONT type 4: Best-effort bandwidth component only, suitable for non delay-
  sensitive bursty traffic.
- TCONT type 5: Any combination of traffic descriptors, suitable for most of the
  general traffic.

Since GPON adopts the strict priority hierarchy in bandwidth allocation, the
received QoS of a request is determined by the TCONT type into which the
request is mapped. The mapping must consider not only the QoS requirement of
the application but also the traffic characteristics [135].

TCONT type 1 is designed to carry delay sensitive CBR traffic. Such traffic
includes legacy TDM voice traffic and rate-controlled broadcast TV and HDTV
streams with CBR characteristics. Another example of an application potentially
mapped into TCONT type 1 is generated by business subscribers, who are willing
to spend more to acquire guaranteed QoS for certain of their applications such as
video conferencing. If these applications are mapped into TCONT type 1, these
applications have a virtually independent channel irrespective of the bandwidth
requests of other applications. Therefore, their QoS can always be guaranteed. This
mapping can also be considered as a scheme for realizing a layer-1 virtual private
network (VPN) for business subscribers.

Fig. 3.5 DBA in GPON

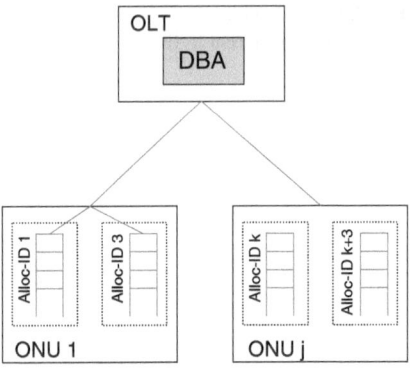

VoIP traffic constructed by an on–off model is an example that constitutes TCONT type 2. The bandwidth during the off period is wasted if it is mapped into TCONT type 1. Thus, it is more bandwidth efficient to map VoIP traffic into TCONT type 2, during which bandwidth is allocated only when there is traffic.

For TCONT type 3, an example is the video on-demand application characterized by the variable bit-rate (VBR) traffic. The assured bandwidth part provisions the application with an average rate guarantee. On the other hand, the VBR characteristic induces either excessive or inadequate bandwidth provision when the application is mapped into TCONT type 1. For the sake of bandwidth efficiency and QoS provisioning, it is proper to map such traffic into TCONT type 3.

For TCONT type 4, best-effort applications, for example, web surfing and file sharing, are a proper fit.

For TCONT type 5, online gaming, which aggregates voice, video, and interactive applications, is a suitable mapping (Fig. 3.5).

3.4   Summary

In this chapter, we have discussed the MAC and resource allocation in GPON systems as standardized in ITU-T G.984 series. In the downstream, the OLT centralizes the traffic multiplexing functionality. The OLT multiplexes GPON frames onto the transmission medium using Port ID as a key to identify the GPON frames that belong to different downstream logical connections. In the upstream direction, the traffic multiplexing functionality is distributed. The OLT grants upstream transmission opportunities, or upstream bandwidth allocations, to the traffic-bearing entities within the subtending ONUs. GPON supports two DBA methods to infer the buffer occupancy status of each ONU: status-reporting DBA and traffic-monitoring DBA. In status-reporting DBA, ONUs report the buffer status to the OLT by directly sending the status report information. In traffic monitoring DBA, the OLT infers the buffer status of ONUs based on the historical information of bandwidth utilization and the amount of assigned bandwidth.

Chapter 4
Media Access Control and Resource Allocation in EPON and 10G-EPON

IEEE802.3ah standardized the MultiPoint Control Protocol (MPCP) as the MAC layer control protocol for EPON [138], and IEEE802.3av specified MAC layer protocols for 10G EPON. For the MAC layer and layers above, in order to achieve backward compatibility such that network operators are encouraged to upgrade their services, 10G-EPON keeps the EPON frame format, MAC layer, MAC control layer, and all the layers above almost unchanged from 1G-EPON [104, 131]. This further implies that similar network management system (NMS), PON-layer operations, administrations, and maintenance (OAM) system, and dynamic bandwidth allocation (DBA) used in EPON can be applied to 10G-EPON as well.

Both IEEE 802.3ah and IEEE 802.3av leave resource allocation in EPON as an open research issue for vendor innovations. Investigating efficient resource allocation algorithms to highly utilize the bandwidth and best arbitrate bandwidth among ONUs as well as queues inside ONUs has thus attracted great research attention in the research community.

4.1 Media Access Control in EPON and 10G-EPON

Figure 4.1 describes the protocol stack of MPCP. In EPON, each ONU in the P2MP topology contains a MPCP entity to communicate with a MPCP entity of the OLT. Thus, a P2P emulation sublayer can be achieved and the P2MP network topology becomes an ensemble of many P2P links for the higher layers.

MPCP describes messages, state machines, and timers to control the channel access of ONUs. Logical Link Identification (LLID) is used to distinguish ONUs in the same PON. LLID is placed in the preamble of each frame, and has a length of two bytes. MPCP performs many functions including auto-discovery, ONU registration, ranging, bandwidth polling, and bandwidth assignment. These functions are performed using a set of 64-byte control messages: GATE, REPORT, REGISTER_REQUEST, REGISTER, and REGISTER_ACK.

N. Ansari and J. Zhang, *Media Access Control and Resource Allocation: For Next Generation Passive Optical Networks*, SpringerBriefs in Applied Sciences and Technology, DOI 10.1007/978-1-4614-3939-4_4, © The Authors 2013

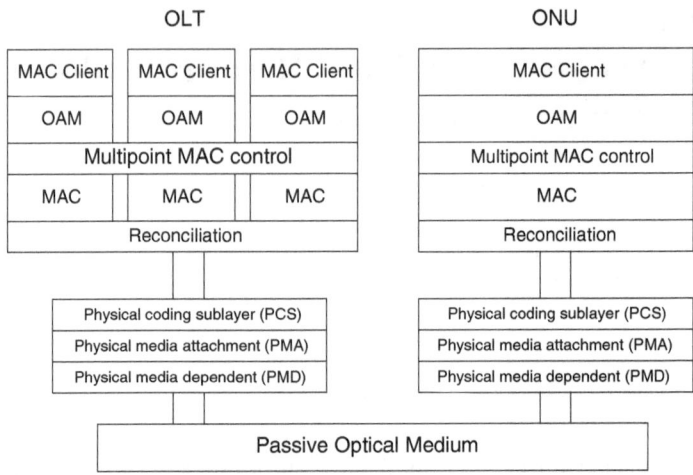

Fig. 4.1 EPON protocol stack

4.1.1 ONU Discovery and Registration

By ONU discovery and registration, newly connected ONUs can be discovered and registered without manual intervention. They can join the EPON system without affecting other ONUs.

The auto discovery and registration process includes the functions of assigning LLID, allocating bandwidth, and compensating different time delays between different ONUs and the OLT. Every time after an ONU is powered on or reset, it waits for the discovery GATE message from the OLT so as to be registered.

Figure 4.2 shows the registration process. The discovery and registration process can be summarized as follows.

- In order to facilitate the discovery and registration of newly connected ONUs, the OLT periodically broadcasts the discovery GATE message to all ONUs. The broadcast message contains a broadcast LLID and multicast destination address.
- The newly connected ONU first waits for the discovery window to open, and then transmits "REGISTER_REQ" message during the opened discovery window. Since multiple newly connected ONUs may transmit "REGISTER_REQ" messages at the same time, each ONU waits for a random amount of time to avoid collisions with the transmission of "REGISTER_REQ" of other ONUs. Once registration collision happens, ONUs will transmit another registration request some time later.
- After the OLT receives the "REGISTER_REQ" messages from ONUs, it sends the "REGISTER" message and assign LLIDs to respective ONUs.
- Following the "REGISTER" message, the OLT sends out a unicast "GATE" message with a unicast LLID and unicast Destination Address (DA) to the ONU.

Fig. 4.2 Registration process
in EPON

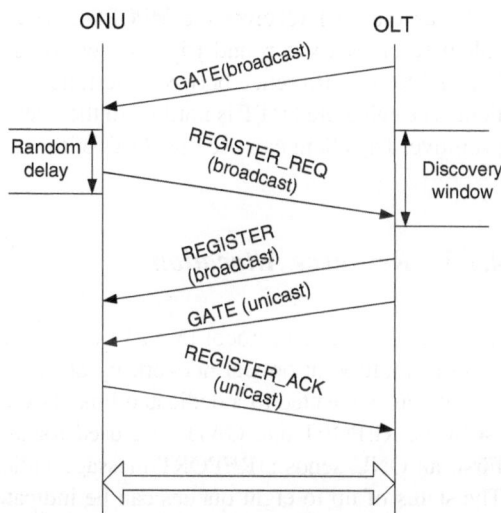

The "GATE" message will be filtered at the regenerator section (RS) layer of the other ONUs and only received by the intended ONU.

- The ONU waits for its initial grant to begin and then sends "REGISTER_ACK" message back to the OLT. The registration process completes.
- After all ONUs are registered, they will transmit frames using a TDMA mechanism which is controlled by the OLT.

4.1.2   Synchronization and Ranging

Time synchronization is critical when using EPON to support time-sensitive services [71]. In EPON, each ONU transmits during the time slots assigned by the OLT. Thus, synchronization between the OLT and the ONU is required to avoid collisions of upstream transmission of ONUs. Every ONU is located at a different distance away from the OLT, and the round trip time (RTT) will change with the changes of time and the environment, and thus transmission overlap of upstream data will likely be incurred.

To avoid the collision of upstream data, RTT between every ONU and the OLT is measured and considered to make the logical distances between the OLT and all the ONUs the same. Both the OLT and the ONU have 32-bit counters that increment every 16 ns. These counters provide a local time stamp. When either device transmits an multipoint control protocol data unit (MPCPDU), it maps its counter value into the timestamp field. The time of transmission of the first octet of the MPCPDU frame from the MAC Control to the MAC is taken as the reference time used for setting the timestamp value. When the ONU receives the MPCPDU, it sets its counter according to the value in the timestamp field in the received MPCPDU.

When the OLT receives the MPCPDU, it uses the received timestamp value to calculate or verify a round trip time between the OLT and the ONU. The RTT is equal to the difference between the timer value and the value in the timestamp field. The calculated RTT is notified to the client via the MA_CONTROL.indication primitive. The client can use this RTT for the ranging process.

4.1.3   Resource Allocation

MultiPoint Control Protocol (MPCP) enables a MAC client to communicate in a point-to-multipoint optical network by allowing it to transmit and receive frames as if it was connected to a dedicated link. Two control messages with the length of 64 bytes, REPORT and GATE, are used for assigning and requesting bandwidth. First, an ONU sends a REPORT message indicating the bandwidth requirements. The status of up to eight queues can be indicated in the REPORT message. After receiving report from an ONU, the OLT can calculate the bandwidth allocation, and then sends out a GATE message to all ONUs. One GATE message can support up to four transmission grants.

Generally, the upstream data transmission is conducted as follows.

- The upstream traffic is queued in the buffer at the ONU side upon arrival.
- The ONU reports the required bandwidth to the OLT using the time slots designated by the OLT.
- The OLT assigns the upstream bandwidth to ONUs, and sends out the result of the bandwidth assignment to ONUs.
- The ONU sends out upstream data to the OLT during the time slot allocated by the OLT.

However, the MPCP protocol does not define any specific bandwidth assignment algorithm. Many DBA algorithms [4, 23, 48, 54, 57] have been developed especially for EPONs to cope with the challenges of high bandwidth utilization and QoS provisioning. However, it is difficult to pick a single best algorithm owing to the multidimensional performance requirements expected of a DBA algorithm. In addition, some algorithms introduce increased complexity when supporting higher traffic demand, QoS, fairness, and so on.

4.2   Dynamic Bandwidth Allocation (DBA)

Neither IEEE 802.3ah nor IEEE 802.3av specifies traffic scheduling algorithms for EPON and 10G-EPON. The bandwidth arbitration among different ONUs is referred to as inter-ONU traffic scheduling, while the bandwidth allocation among queues in the same ONU is referred to as intra-ONU traffic scheduling [73]. Figure 4.3 describes inter-ONU scheduling and intra-ONU scheduling, respectively.

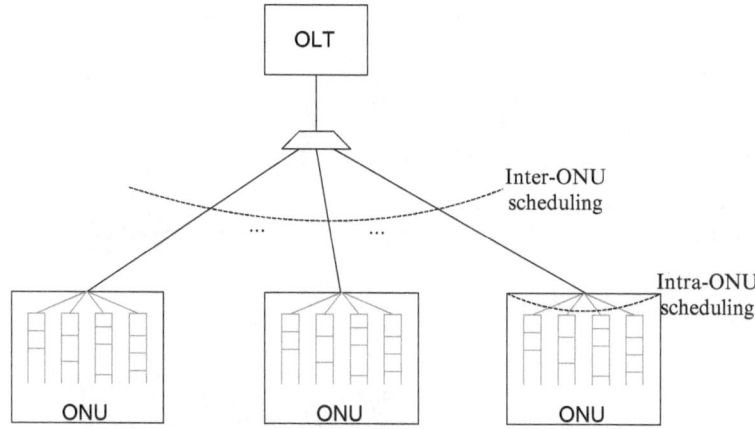

Fig. 4.3 Inter-ONU scheduling and intra-ONU scheduling

Generally, the objective of the traffic scheduling scheme is to best utilize the bandwidth and guarantee different requirements of queues in all ONUs. Thus, bandwidth utilization, QoS control, and fairness assurance constitute three main objectives in designing DBA algorithms.

4.2.1 Inter-ONU Bandwidth Allocation

4.2.1.1 Bandwidth Utilization

One intuitive bandwidth allocation scheme is to allocate a fixed amount of bandwidth to each ONU disregarding the actual incoming traffic. However, traffic of ONUs are rather dynamic, and adopting this method may result in significant bandwidth under utilization. High bandwidth utilization requires adapting the bandwidth allocation to the real-time incoming user traffic.

In order to adapt the bandwidth allocation to the real-time arrival traffic, the upstream queue status of an ONU needs to be polled before assigning the bandwidth to the ONU. Figure 4.4 illustrates the polling and upstream traffic transmission procedure. The OLT first polls the upstream queue status of ONU i. After receiving the response from ONU i, the OLT allocates the bandwidth to ONU i. Polling of ONU $i+1$ happens after receiving the last transmission bit of ONU i. However, this scheme introduces a time gap between traffic transmission of two ONUs. As indicated in Fig. 4.4, the time gap is at least two round-trip times (RTTs) between the ONU and the OLT. To better utilize the bandwidth, Kramer et al. [57] proposed interleaved polling with adaptive cycle time (IPACT).

As a seminal work in EPON DBA, IPACT interleaves polling messages with Ethernet frame transmission, as shown in Fig. 4.5. It reduces the overhead arisen from

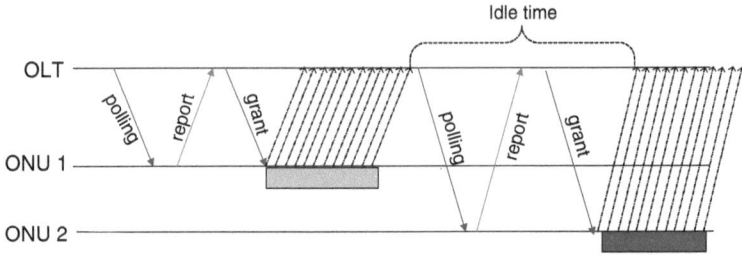

Fig. 4.4 An example of the DBA

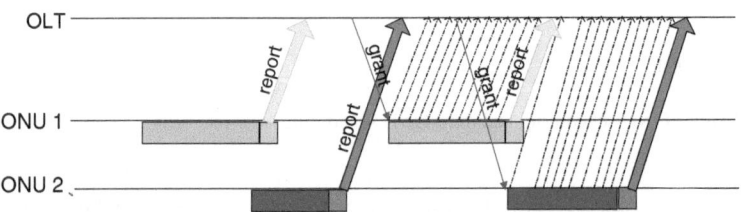

Fig. 4.5 IPACT

the different round-trip times (RTTs) of ONUs in the polling system. Furthermore, the polling cycle adapts to the traffic demand of ONUs that maximizes the statistical multiplexing gain as opposed to static TDM systems.

In IPACT, rather than polling the $i+1$th ONU after receiving the last bit of ONU i, ONU $i + 1$ is polled and granted with bandwidth while ONU i is still transmitting the traffic as depicted in Fig. 4.5. Also, when the ith ONU is transmitting Ethernet frames in the upstream, the OLT informs ONU $i + 1$ about the grant information, including the starting time and the size of the granted bandwidth. The bandwidth allocation scheme guarantees that the first bit from ONU $i + 1$ arrives at the OLT only after the guard time has passed (i.e., after the OLT receives the last bit from the ith ONU). In addition, two basic requirements are to be met:

- The GATE message carrying the grant information should arrive at the $i + 1$th ONU in time.
- The bandwidth granted for the ith ONU is equal to the bandwidth requested by the ith ONU.

However, in this scheme, an ONU reports its bandwidth to the OLT immediately after it sends out upstream traffic. By the time that the OLT allocates bandwidth to the ONU, the information of the queue status may be outdated already, thus resulting in the large delay of the traffic which arrives between the last report and the current grant.

4.2.1.2 Delay

To further reduce the packet delay, and better guarantee QoS for users, prediction based traffic scheduling schemes have been proposed. In the prediction based schemes, traffic which arrives after sending reports to the OLT in the last cycle is estimated, and allocated with some bandwidth [69, 70]. The traffic prediction is made according to the historical traffic arrival information based on self-similarity of the network traffic.

The prediction scheme works well for periodic constant bit rate (CBR) traffic, e.g., traditional voice traffic. However, under-utilization may be resulted when the network traffic is bursty. For further studies, readers are referred to a rigorous investigation on applying nonlinear predictor-based dynamic resource allocation schemes to improving the upstream transmission efficiency of a P2MP network [120]. Using PON as an example of a P2MP network, a general state space model was established to analyze the controllability and stability of the P2Mp network from the system point of view. Based on control theory, Yin and Ansari [120] provided guidelines to design a controller to maintain the system stability as well as to design an optimal compensator to achieve system accuracy.

4.2.1.3 Fairness

In deciding the amount of bandwidth allocated to ONUs, the OLT may grant an ONU with the amount of bandwidth requested by the ONU. However, ONUs with large bandwidth requirements may monopolize the bandwidth usage. Consequently, ONUs with small bandwidth demands may experience large delays. To ensure fairness among ONUs, some rules are set to upper bound the bandwidth granted to each ONU in a cycle.

IPACT-LS prevents ONUs from monopolizing the bandwidth by setting a predetermined maximum of the granted resources [58]. Assi et al. [4] proposed to satisfy requests from light-load ONUs first while penalizing heavily-loaded ONUs. Typically, there are three ways to limit the bandwidth allocated to an ONU.

- Limited service: This DBA algorithm grants the requested number of bytes, but no more than the Maximum Transmission Window (MTW).
- Credit service: This DBA algorithm grants the requested window plus either a constant credit or a credit that is proportional to the requested window.
- Elastic service: This DBA algorithm attempts to overcome the limitation of assigning at most one fixed MTW to an ONU in a cycle. The maximum window granted to an ONU is such that the accumulated size of the last N grants does not exceed N MTWs, where N denotes the number of ONUs. Thus, if only one ONU is backlogged, it may get a grant of up to N MTWs.

4.2.2   Intra-ONU Scheduling

4.2.2.1   QoS

QoS control plays an important role especially when the traffic load increases beyond the network capacity such that QoS requirements of queues have to be sacrificed. Considering different QoS requirements of applications, many strategies have been proposed to arbitrate bandwidth among queues in the same ONU. Kramer et al. [56] and Assi et al. [4] incorporated the DiffServ framework into IPACT. DiffServ classifies the incoming traffic into three classes: the expedited forwarding (EF), the assured forwarding (AF), and the best effort (BE). EF services gather the delay sensitive applications that require a bounded end-to-end delay and jitter specifications, whereas the AF class is intended for services that are not delay sensitive but require bandwidth guarantees. Finally, BE services are not delay sensitive and have neither jitter specifications nor minimum guaranteed bandwidth. Kramer et al. [56] and Luo and Ansari [69] proposed to provision EF and AF traffic with higher priority than AF and BE traffic, respectively. Bai et al. [5] proposed to divide the DBA cycle into two subcycles. The first subcycle is allocated to EF traffic whereas the second one is for AF and BE traffic.

Assigning queues with priorities in intra-ONU scheduling raises the fairness issue. Always giving one queue with the highest priority may be unfair to the other queues. For example, incorporating the DiffServ framework into EPON DBA with the strict-priority discipline causes the so-called *light-load penalty* problem [56]. Owing to the strict priority, the newly arriving high-priority traffic after sending REPORT will preempt the granted bandwidth for reported low-priority traffic, thus resulting in the starvation and unfairness of low-priority queue.

4.2.2.2   Fairness

To ensure fairness among ONUs, Kramer et al. [56] further proposed a two-stage queuing system, where packets arriving before sending REPORT are queued in the first-stage, and those arriving after sending REPORT are queued in the second-stage. The light-load penalty is compensated by setting the rule that the high priority traffic queued in the second stage are not allowed to preempt the transmission of the low-priority traffic.

Kramer et al. [56] proposed to use a proper local queue management scheme and proposed priority-based scheduling to compensate for the light load penalty. So, packets arriving before a time stamp with highest priority are scheduled first, and packets arriving past the time stamp are then scheduled. Many other works have been proposed to tackle the intra-ONU scheduling issue. For example, References [54, 82] adopted weighted fair queuing to assign queues with different weights according to their priorities. Ghani et al. [23] used a virtual-scheduler to schedule queues.

4.3 Maximizing User QoE in DBA

Next generation network services should be delivered in the most cost- and resource-efficient manner with ensured user satisfaction [127]. Thus, service providers are driving to provision user quality of experience (QoE), which indicates the overall network performance with respect to the user satisfaction. Here, we present a QoE-oriented dynamic bandwidth allocation scheme for a PON system. The OLT is assumed to have virtual output queues for the downstream traffic. Each of these virtual output queues corresponds to one user session at one ONU. Upon the arrival of the downstream traffic of each user session, the OLT arbitrates the bandwidth allocation among different user sessions at different ONUs, and then dispatches the arrival traffic.

Consider deterministic entry of the downstream traffic into the network. Denote $\{\mathbf{r}_{i,j}^{k}\}_{i,j,k}$ as the kth request of session j at ONU i. Each request corresponds to some packets, which arrive during a continuous time duration. $\mathbf{r}_{i,j}^{k}$ is associated with a double sequence $(a_{i,j}^{k}, x_{i,j}^{k})$, where $x_{i,j}^{k}$ is the size expressed in time duration of request $\mathbf{r}_{i,j}^{k}$, and $a_{i,j}^{k}$ is the time at which request $\mathbf{r}_{i,j}^{k}$ arrives. For the downstream case, both $\{x_{i,j}^{k}\}_{i,j,k}$ and $\{a_{i,j}^{k}\}_{i,j,k}$ are known to the decision maker OLT.

For the upstream scenario, upstream traffic arrives at ONUs, and is not explicitly known to the OLT. Since the decision maker OLT does not know the exact arrival time and size of upstream arrival traffic, the upstream scenario in PONs is more complicated than the downstream scenario. Different from the downstream scenario where the OLT can closely keep track of the traffic arrival information, the OLT, in the upstream scenario, does not directly own the exact information of the traffic arrival time and size, i.e., $(a_{i,j}^{k}, x_{i,j}^{k})$, which needs to be estimated based on ONU reports.

4.3.1 QoE

4.3.1.1 QoE of an User

We identify QoE of a user based on the performance of the session with the lowest QoE score among all sessions of the user. That is to say, QoE of user i equals to $\min_j u_{i,j}$, where $u_{i,j}$ refers to QoE of session j of ONU i, and it can be regarded as the QoE score of user i when user i has session j only. Achieving a given QoE score v for user i implies that

$$u_{i,j} \geq v, \forall j \tag{4.1}$$

4.3.1.2　QoE of an User Session

QoE of a session achieves QoE score v only when its loss and delay satisfy certain criteria. Denote $loss_{i,j}^{-1}(v)$ and $delay_{i,j}^{-1}(v)$ as the maximum allowable loss and delay for session j at ONU i to achieve QoE score v, respectively, i.e.,

$$\begin{cases} loss_{i,j}^{-1}(v) = \arg\max_{loss}\{u_{i,j} = v\} \\ delay_{i,j}^{-1}(v) = \arg\max_{delay}\{u_{i,j} = v\} \end{cases}$$

Then, the necessary and sufficient condition to guarantee QoE score v for user i is that

$$l_{i,j} \leq loss_{i,j}^{-1}(v) \text{ and } d_{i,j} \leq delay_{i,j}^{-1}(v), \forall j \tag{4.2}$$

where $l_{i,j}$ and $d_{i,j}$ denote the loss and delay performances of session j at ONU i, respectively.

We assume QoE functions $u_{i,j}$, $loss_{i,j}^{-1}$, and $delay_{i,j}^{-1}$ are known a priori [123, 124]. Investigating these QoE functions is rather challenging, and has received intensive research attention [19, 36, 45, 91, 97, 105].

4.3.1.3　Delay and Loss

The subsequent question is how to define loss ratio and delay for user sessions. While many proposed works employ the average loss ratio as the loss metric, the application-level QoS perceived by end users is also affected by short-term loss patterns (loss burstiness and loss interval) [92, 98]. For delay, we expect that each traffic request has a bounded delay such that the delivery of this particular request is on time without degrading user experience. The short term delay can also guarantee delay jitter performance [109]. Thereby, we assume that $l_{i,j} \leq loss_{i,j}^{-1}(v)$ and $d_{i,j} \leq delay_{i,j}^{-1}(v)$, further implying that delay and loss of every single request in the session are less than $loss_{i,j}^{-1}(v)$ and $delay_{i,j}^{-1}(v)$, respectively.

Denote $c_{i,j}^k$ and $y_{i,j}^k$ as the completion transmission time and the granted time duration of the kth request of session j at user i (request $\mathbf{r}_{i,j}^k$), respectively. Define traffic loss ratio of request $\mathbf{r}_{i,j}^k$ as $(x_{i,j}^k - y_{i,j}^k)/x_{i,j}^k$, and the delay of request $\mathbf{r}_{i,j}^k$ as the difference between the request completion time and the request arrival time, i.e., $c_{i,j}^k - a_{i,j}^k$.

Then, mathematically, guaranteeing QoE score v of session j at user i implies that

$$\begin{cases} (x_{i,j}^k - y_{i,j}^k)/x_{i,j}^k \leq loss_{i,j}^{-1}(v), \forall k \\ c_{i,j}^k - a_{i,j}^k \leq delay_{i,j}^{-1}(v), \forall k \end{cases} \tag{4.3}$$

Further, guaranteeing QoE score v of user i implies that

$$
\begin{cases}
(x_{i,j}^k - y_{i,j}^k)/x_{i,j}^k \leq loss_{i,j}^{-1}(v), \forall k, \forall j \\
c_{i,j}^k - a_{i,j}^k \leq delay_{i,j}^{-1}(v), \forall k, \forall j
\end{cases}
\tag{4.4}
$$

For ease of explanation, for any request $\mathbf{r}_{i,j}^k$, the largest QoE score v satisfying $(x_{i,j}^k - y_{i,j}^k)/x_{i,j}^k \leq loss_{i,j}^{-1}(v)$ and $c_{i,j}^k - a_{i,j}^k \leq delay_{i,j}^{-1}(v)$ is referred to as QoE of request $\mathbf{r}_{i,j}^k$ for the rest of the chapter.

4.3.1.4 Problem Formulation

Out of fairness concern, we try to achieve *max–min fairness* among QoE of all users in allocating resources [81, 93]. With the above definitions and assumptions, the problem of achieving max–min fairness can be formulated as:

Given traffic requests $\{\mathbf{r}_{i,j}^k\}$, QoE functions $u_{i,j}$, $loss_{i,j}^{-1}$, and $delay_{i,j}^{-1}$, $\forall i, \forall j$, construct a schedule with the smallest delay $l_{i,j}^k$ and loss $d_{i,j}^k$ for all requests such that the QoE score of any ONU i cannot be increased at the sacrifice of the decrease of the QoE score of any other ONU whose QoE is already smaller than that of ONU i.

We first focus on the downstream bandwidth allocation problem in which the decision maker OLT can track $\{(a_{i,j}^k, r_{i,j}^k)\}$ of all requests. Subsequently, we discuss the estimation of $\{(a_{i,j}^k, r_{i,j}^k)\}$ in the upstream scenario.

4.3.2 The Downstream Scenario

We assume that downstream traffic arrivals during the whole time span, i.e., $\{(a_{i,j}^k, r_{i,j}^k)\}_{k=1}^{\infty}, \forall i, \forall j$, are known in advance. Then, inspired from the optimal scheduling scheme described in the above, we present a scheduling scheme for real-time implementation in which the decision maker does not know the future incoming traffic at the decision making time.

4.3.2.1 Maximize the Minimum QoE of All Users

We first look at the problem of achieving a given QoE score for all users. For a given QoE score v, based on Constraints (4.4), guaranteeing v for all users implies that

$$
\begin{cases}
(x_{i,j}^k - y_{i,j}^k)/x_{i,j}^k \leq loss_{i,j}^{-1}(v), \forall i, \forall j, \forall k \\
c_{i,j}^k - a_{i,j}^k \leq delay_{i,j}^{-1}(v), \forall i, \forall j, \forall k
\end{cases}
\tag{4.5}
$$

Algorithm 1 Guarantee QoE score v for all users

1: Consider $x_{i,j}^k \cdot (1 - loss_{i,j}^{-1}(v))$ as the grant size $y_{i,j}^k$ for request $\mathbf{r}_{i,j}^k$.
2: Consider $a_{i,j}^k + delay_{i,j}^{-1}(v)$ as the scheduling deadline of request $\mathbf{r}_{i,j}^k$.
3: $t = 0$
4: **while** There exists unscheduled request **do**
5:    Among unscheduled request $\mathbf{r}_{i,j}^k$ which arrive before t, select the one with the earliest deadline.
6:    **if** request $\mathbf{r}_{i',j'}^{k'}$ arrives between t and $t + y_{i,j}^k$ and has deadline earlier than request $\mathbf{r}_{i,j}^k$ **then**
7:       $c_{i,j}^k = a_{i',j'}^{k'}$
8:    **else**
9:       $c_{i,j}^k = t + y_{i,j}^k$ and denote request $\mathbf{r}_{i,j}^k$ as scheduled.
10:    **end if**
11:    Allocate the time between t and $c_{i,j}^k$ to request $\mathbf{r}_{i,j}^k$.
12:    $t = c_{i,j}^k$
13: **end while**

Then, the problem is equivalent to the problem of constructing a schedule with $y_{i,j}^k$ and $c_{i,j}^k$ satisfying $y_{i,j}^k \geq x_{i,j}^k \cdot (1 - loss_{i,j}^{-1}(v))$ and $c_{i,j}^k \leq a_{i,j}^k + delay_{i,j}^{-1}(v), \forall i, \forall j, \forall k$. The following Algorithm 1 addresses this problem.

Algorithm 1 is essentially a preemptive earliest-deadline-first (EDF) scheduling algorithm. Among all unscheduled requests, the one with the earliest deadline is scheduled with the highest priority. When a request is being scheduled, the scheduling can be preempted and resumed later if another request with an earlier deadline arrives.

Theorem. *If the schedule constructed by Algorithm 1 cannot guarantee all users with the QoE score v, then, no schedule exists to guarantee all users with the QoE score v.*

Proof. Assume the deadline of request $\mathbf{r}_{i,j}^k$ is violated in the schedule constructed by Algorithm 1, i.e., $c_{i,j}^k > a_{i,j}^k + delay_{i,j}^{-1}(v)$. We show that there does not exist a schedule which can schedule all requests prior to their respective deadlines.

It is not difficult to see that the scheduling policy described in Algorithm 1 is work-conservative. Then, the earliest time to schedule all requests with deadline earlier than $a_{i,j}^k + delay_{i,j}^{-1}(v)$ is $c_{i,j}^k$. If request $\mathbf{r}_{i,j}^k$ is scheduled earlier, there must exist some other request $\mathbf{r}_{i',j'}^{k'}$ with deadline earlier than $a_{i,j}^k + delay_{i,j}^{-1}(v)$ that completes its transmission at time $c_{i,j}^k$. In this case, the deadline request $\mathbf{r}_{i',j'}^{k'}$ is violated.   □

If there exists a schedule guaranteeing the QoE score v for all users, we say v is achievable for all users. With the problem of guaranteeing the QoE score v for all users being addressed, the minimum achievable QoE score for all users can be maximized by trying different v using the bisection method [11] which is described in Algorithm 2.

Algorithm 2 Maximize the minimum QoE among all users by employing the bisection method

---

1: Denote h and l as the highest and lowest value of QoE functions of all applications, respectively.
2: $v = h$
3: **while** h and l are not close enough **do**
4:    **if** v is achievable for all users **then**
5:       $l = v$
6:    **else**
7:       $h = v$
8:    **end if**
9:    $v = (h+l)/2$
10: **end while**

Algorithm 3 Further increase QoE of some user sessions

1: **while** QoE score can be increased for some session **do**
2:    Decide the sessions whose QoE can be possibly increased, i.e., sessions which have not reached their respective highest QoE scores yet.
3:    Maximize the minimum QoE score for these sessions.
4: **end while**

The main idea is as follows: Let h and l be the highest and lowest value of QoE functions of all sessions. We first let v equal to h, and check whether v is achievable for all users. If v is not achievable, h is updated to be v, and v is decreased to the midpoint between h and l; otherwise, h is increased to v, and v is increased to the midpoint between h and l. The above process is performed recursively until h and l are close enough to each other.

4.3.2.2 Further Increase of QoE of Some Sessions if Possible

In the above, the schedule with the maximum achievable QoE score for all users is obtained. Although there does not exist a better schedule which can increase the QoE score of all sessions of all users at the same time, the QoE score of some sessions of some users may be increased without decreasing those of other sessions. For example, assume user sessions can be classified into two classes, in which the two classes with the highest QoE scores are μ and v, respectively. Without loss of generality, assume $\mu < v$. Then, using Algorithm 2, a QoE score higher than μ cannot be achieved for sessions in the second class whose QoE score can be as high as v.

To further increase QoE scores of some sessions, we propose Algorithm 3, which is similar to the idea of water-filling. In Algorithm 3, we consider all sessions whose QoE scores have not reached their respective highest scores, and try to maximize the minimum QoE of all these sessions. The process is repeated until the QoE score cannot be further increased for any session. The detailed procedure involved in Line 3 of Algorithm 3 is similar to that described in Algorithm 2.

Algorithm 4 Further increase QoE scores of some requests

1: /*Enhance delay and loss performances for Case 1*/
2: **while** QoE score can be increased for some request **do**
3:    Among all requests of all sessions, decide requests whose QoE scores cannot be further increased.
4:    Obtain the maximum QoE score which can be guaranteed for all the other requests.
5: **end while**
6: /*Enhance delay and loss performances for Case 2*/
7: Obtain requests being scheduled with the highest QoE score.
8: Gradually decrease the loss ratio of these requests until no further loss can be made.
9: Obtain requests being scheduled with zero loss ratio.
10: Gradually decrease the delay of these requests until no further delay can be made.

4.3.2.3 Further Improvement of Delay and Loss of Some Requests

Using Algorithms 2 and 3, we have constructed a schedule with which QoE of any session i cannot be increased without the sacrifice of QoE of some other sessions whose QoE scores are already smaller than that of session i. Although QoE of a session cannot be increased, delay and loss of some requests in the session may be enhanced for two cases. First, the fact that QoE of a session cannot be increased implies that delay and loss performances of all requests in the session cannot be decreased simultaneously. However, for some requests in a session, their delay and loss performances may be possibly enhanced. This case is referred to as *Case 1*. Second, to achieve the highest QoE score, usually, delay and loss do not need to be as small as zero owing to the human auditory and visual limitation. In the schedule produced by Algorithms 2 and 3, if the highest QoE score of a session is achieved, delay and loss ratio of requests in the session equals to the maximum allowable value to achieve the highest QoE score. It is possible to further reduce delay and loss ratio from the maximum allowable value for some requests, or even all requests of a session. This case of further reducing delay and loss is referred to as *Case 2*.

Algorithm 4 describes the scheme of further enhancing delay and loss performances of some requests. In the real implementation, delaying the scheduling of a request is much easier than dropping some traffic of the request. Thus, delaying traffic is preferable over dropping traffic when either one has to be chosen. In Algorithm 4, delay and loss performances are first enhanced for requests in Case 1. In the "while" loop described in Lines 2–5 of Algorithm 4, we maximize the minimum QoE score of requests which can be possibly increased. Then, the requests whose QoE scores can be increased are updated, and the same process repeats until no request can achieve an increased QoE score without degrading QoE of others. In Lines 7–8 of Algorithm 4, the algorithm first obtains all requests which achieve their respective highest QoE scores. Then, the algorithm tries to gradually reduce the loss ratio of these requests until no further loss can be made. In Lines 9–10 of Algorithm 4, we first obtain all requests with zero loss ratio, and then try to gradually decrease delay of these requests until no further delay can be made.

4.3.3   The Upstream Scenario

Different from the downstream scenario, the bandwidth allocation decision maker OLT does not own the exact information of the arrival time $a^k_{i,j}$ and size $x^k_{i,j}$ of each request. In this section, we discuss the estimation of $a^k_{i,j}$ and $x^k_{i,j}$ in the upstream scenario.

We first introduce several more notations to facilitate the estimation of the arrival upstream requests. Denote $\alpha^k_{i,j}$ as the time of the kth report of session j generated by ONU i. Then, at time $\alpha^k_{i,j} + RTT_i/2$, the OLT receives the kth report of session j at ONU i, where RTT_i is the round trip time between ONU i and the OLT. Denote $\Delta^k_{i,j}$ as the interval between the $(k-1)$th and kth request sending time from session j of ONU i, i.e., $\Delta^k_{i,j} = [\alpha^{k-1}_{i,j}, \alpha^k_{i,j}]$. Denote $tr^k_{i,j}$, $dr^k_{i,j}$, and $ar^k_{i,j}$ as the transmitted traffic, dropped traffic, and arrival traffic during interval $\Delta^k_{i,j}$, respectively. Then, at time $\alpha^k_{i,j}$, the backlogged traffic of session j at ONU i equals to

$$\sum_{p=1}^{k} \left(ar^p_{i,j} - tr^p_{i,j} - dr^p_{i,j} \right)$$

Denote $\gamma^k_{i,j}$ as the reported queue length (in time) contained in the kth report of session j at ONU i. Then, $\gamma^k_{i,j} = \sum_{p=1}^{k} \left(ar^p_{i,j} - tr^p_{i,j} - dr^p_{i,j} \right)$. We can further obtain

$$ar^k_{i,j} = \gamma^k_{i,j} + \sum_{p=1}^{k} \left(tr^p_{i,j} + dr^p_{i,j} \right) - \sum_{p=1}^{k-1} ar^p_{i,j} \tag{4.6}$$

At the right side of Eq. (4.6), $\gamma^k_{i,j}$ is the request reported to the access node. Both $tr^p_{i,j}$ and $dr^p_{i,j}$ are decided by the OLT. Hence, by recursion, the OLT can infer the arrival traffic $ar^k_{i,j}$ during time interval $\Delta^k_{i,j}$.

Besides the newly arrival traffic $ar^k_{i,j}$ during interval $\Delta^k_{i,j}$, the OLT can estimate the arrival time of all backlogged traffic at time $\alpha^k_{i,j}$. We assume infinite buffer size at the user side, and the dropping is from the head of the queue. As compared to the scheme of dropping the latest arrival packets, dropping the oldest packets first can let precious resources be used for transmitting traffic with smaller delay, and hence larger QoE.

Then, among the total $\sum_{p=1}^{k} ar^p_{i,j}$ arrival traffic before time $\mathbf{a}^k_{i,j}$, the first $\sum_{p=1}^{k} (tr^p_{i,j} + dr^p_{i,j})$ arrival traffic is either transmitted or dropped, and the latest arrival $\sum_{p=1}^{k} (ar^p_{i,j} - tr^p_{i,j} - dr^p_{i,j})$ traffic remains in the queue and is reported to the central access node. Among the $\gamma^k_{i,j}$ request traffic, assume $\eta^k_{i,j}(p)$ traffic arrives during time interval $\Delta^p_{i,j}$. Then, we can obtain that

$$\begin{cases} \eta_{i,j}^k(k) = \min\{ar_{i,j}^k, \gamma_{i,j}^k\} \\ \eta_{i,j}^k(p) = \min\left\{ar_{i,j}^p, \left(\gamma_{i,j}^k - \sum_{m=l+1}^{k} \eta_{i,j}^{k,m}\right)^+\right\} \end{cases} \quad (4.7)$$

where $x^+ = \begin{cases} x \text{ if } x > 0 \\ 0 \text{ otherwise} \end{cases}$.

After the arrival time and size of each request are estimated, the bandwidth allocation problem is boiled down to the problem discussed in the downstream scenario.

4.3.4　Simulation Results and Analysis

In this section, taking the upstream transmission in EPON for example, we investigate the performance of our proposed online scheduling scheme.

IEEE 802.3ah has standardized the MultiPoint Control Protocol (MPCP) as the MAC layer control protocol for EPON [69]. Specifically, MPCP defines two 64-byte control messages REPORT and GATE for the bandwidth arbitration in the upstream. ONUs report its backlogged traffic to the OLT by sending REPORT. After collecting REPORT from ONUs, the OLT dynamically allocates bandwidth to ONUs and informs its grant decisions to ONUs via GATE. The cycle duration in EPON can be dynamically adaptive to the traffic. In the simulation, instead of letting the OLT calculate schedules every time a REPORT from an ONU arrives, the OLT keeps collecting REPORTs from different ONUs, and makes bandwidth allocation decision just before the upstream wavelength channel becomes idle.

The interval between two consecutive bandwidth allocation decision making time is referred to as a dynamic bandwidth allocation (DBA) cycle in this chapter. The DBA cycle is adapted to traffic variation. We set the maximum DBA cycle as 2 ms, and the data rate as 1.25 Gb/s. Then, the maximum traffic transmitted during a cycle is 2.5 Mbits. When the total requests are below 2.5 Mbits, every request is granted with the bandwidth equaling to its request size. When the total requests increase beyond 2.5 Mbits, the OLT allocates the bandwidth in the cycle with the duration of 2 ms to ONUs.

For ease of explanation, we define three kinds of DBA cycles according to the traffic load. For one particular DBA cycle, if the total requested traffic is within the capacity of the cycle, the cycle is referred to as a *low-load cycle*; if the requested traffic is greater than the capacity of the cycle but every ONU can still get the highest QoE score by delaying the scheduling of some traffic to the next cycle or dropping some traffic, the cycle is referred to as a *medium-load cycle*; if the QoE score of some ONUs falls below the highest score, the cycle is referred to as a *high-load cycle*.

We set the number of ONUs to be 16, and the round trip time between each ONU and the OLT to be 125 μs. The simulation model is developed on the OPNET platform. Since self-similarity is exhibited in many applications, we input each user

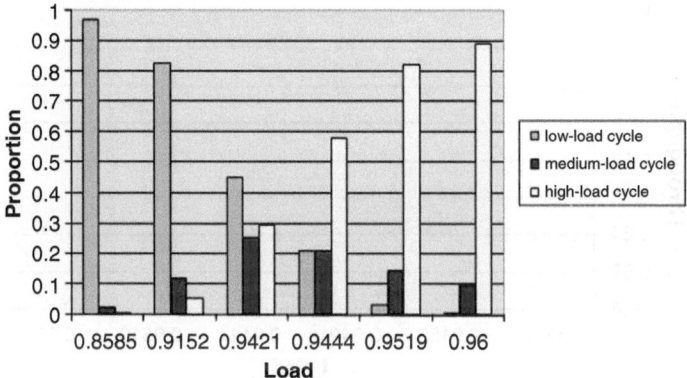

Fig. 4.6 The proportion of three kinds of cycles under different traffic loads

session with self-similar traffic. The Pareto parameter is set as 0.8. The packet length is uniformly distributed between 64 bytes and 1,500 bytes. We assume each ONU has five sessions corresponding to five kinds of applications. Each of the five sessions is entered with traffic with the same statistical characteristics. The input traffic of all ONUs obey the same distribution.

First, we investigate distributions of low-load cycles, medium-load cycles, and high-load cycles under different traffic loads. The simulation time is set to be 2.5 s. We assume the maximum allowable delay to achieve the highest QoE score of the fiver sessions in each ONU are 3 ms, 4 ms, 5 ms, 6 ms, and 7 ms, respectively, no traffic loss is allowed, and the precise QoE functions are unknown. Figure 4.6 presents simulation results of the distribution of the three kinds of cycles. It was shown that when the network load is about 0.8585, the majority of the cycles are low-load cycles, and the high-load cycles are very few. That is to say, the maximum QoE can almost be guaranteed under this case. When the network load is increased to 0.9152, the medium-load cycles are increased to around 11% of the total number of cycles, and the high-load cycles assume around 5% of the total number of cycles. Then, in 5% of all cycles, QoE of some sessions are degraded to some degree. When the network load increases to 0.96, the majority of cycles are high-load cycles, implying that tremendous amount of computation is required.

Figure 4.7 shows the throughput under different traffic loads; the traffic load is defined as the ratio of the sum of input traffic during the whole simulation time over the maximum traffic accommodated by the network. Throughput is defined as the ratio of the total amount of successfully transmitted traffic to the maximum traffic which can be accommodated by the network. Under low traffic load, all the requests can be successfully scheduled. Hence, throughput increases with the increase of the traffic load. When the traffic load is increased to a certain value, further increase will not increase the throughput of the network. As shown in Fig. 4.7, the knee point happens when the network load is around 0.9421, where high-load cycles occupy

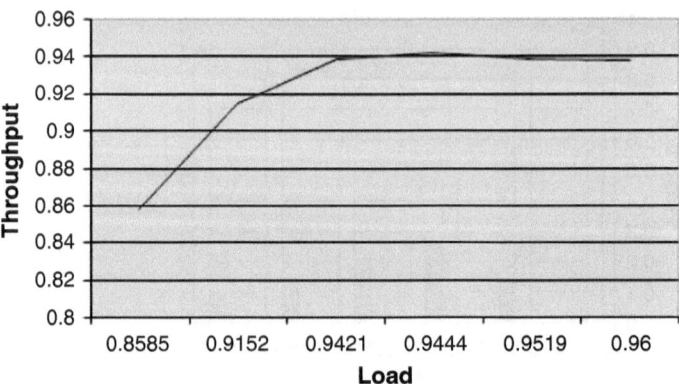

Fig. 4.7 Throughput vs. traffic load

less than 30% of the total cycles. In another words, most of the cycles will be either low-load cycles or medium-load cycles, which do not involve much computation, and more importantly, the highest QoE score can be guaranteed in these low-load and medium load cycles.

We next assume QoE functions are known, and consider two kinds of QoE functions. One is a function of packet delay, and the other one is a function of loss ratio.

4.3.4.1 Packet Delay

First, we consider QoE of a user session as a function of packet delay, i.e., $u_{i,j}(loss, delay) = u_{i,j}^2(delay)$, defined as follows.

$$u_{i,1}^2(delay) = \begin{cases} 1 & delay \leq 3\,\text{ms} \\ e^{(delay-3)/3} & delay > 3\,\text{ms} \end{cases}, \forall i$$

$$u_{i,2}^2(delay) = \begin{cases} 1 & delay \leq 4\,\text{ms} \\ e^{(delay-4)/4} & delay > 4\,\text{ms} \end{cases}, \forall i$$

$$u_{i,3}^2(delay) = \begin{cases} 1 & delay \leq 5\,\text{ms} \\ e^{(delay-5)/5} & delay > 5\,\text{ms} \end{cases}, \forall i$$

$$u_{i,4}^2(delay) = \begin{cases} 1 & delay \leq 6\,\text{ms} \\ e^{(delay-6)/6} & delay > 6\,\text{ms} \end{cases}, \forall i$$

$$u_{i,5}^2(delay) = \begin{cases} 1 & delay \leq 7\,\text{ms} \\ e^{(delay-7)/7} & delay > 7\,\text{ms} \end{cases}, \forall i$$

We assume the traffic is delayed if it fails to be successfully transmitted. The buffer size of each queue is set as 25 K bytes to avoid queue build-up at high loads. The network load is defined as the ratio of the total traffic admitted into the network to the capacity of the network.

Then, we want to show that QoS profiles received by the five kinds of sessions conform to the corresponding profiles derived from their application utilities. Fairness is achieved if application utilities obtained by sessions are equivalent to each other.

Figure 4.8 shows the average delay of packets received in cycles under six different network loads. Since the input traffic is self-similar, the delay of packets fluctuates cycle from cycle. Generally, with the increase of the traffic load, the packet delay increases. When the network load equals to 0.8585, most of the cycles are low-load cycles. Hence, almost all requests from these five kinds of sessions can be scheduled immediately. The five sessions experience similar delay performance. When the network load equals to 0.9162, around 15% cycles are medium-load cycles or high-load cycles, where requests from some of the sessions are delayed. It is shown that session 5 experiences the largest delay while session 1 has the smallest delay, complying to their respective QoE profiles. With the increase of traffic load, delay of sessions increases but at different degrees, as determined by their respective application utilities. Delay of session 1 increases at the smallest degree, while that of session 5 increases at the largest degree. When the traffic load increases to 0.96, most of the cycles are high-load cycles. This implies that most of the requests have to be delayed before being successfully transmitted. Simulations show that delay in this case is much higher than that in the case when the load is 0.8585.

Figure 4.9 shows QoE of the five kinds of sessions under different loads. It can be seen that QoE of all sessions are almost the same with small differences under a particular traffic load. For each session, there exists obvious differences in delay with different traffic loads. However, the difference in QoE is not that obvious, i.e., very small. The average QoE achieved when the load equals to 0.96 is slightly lower than that achieved when the load is 0.8585. This can be attributed to the fact that QoE is set as the same value when the delay is below a certain value.

4.3.4.2 Loss Ratio

Next, we consider QoE as a function of packet loss ratio, i.e., $u_{i,j}(loss, delay) = u^1_{i,j}(loss)$. In a particular cycle, if the request is greater than the capacity of the cycle, the extra requested traffic is dropped rather than delayed. The buffer size for each queue for a user session is set as infinity. For the five sessions in each ONU, $u^1_{i,j}(loss)$ is defined as follows.

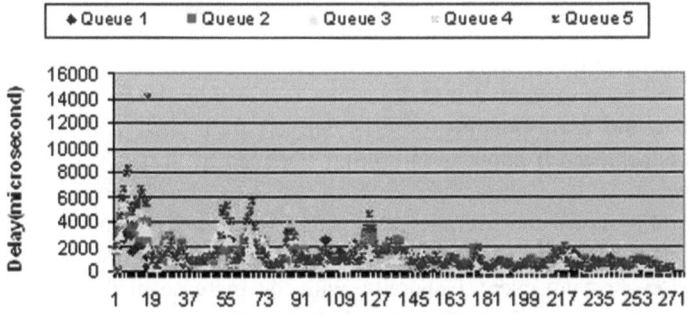

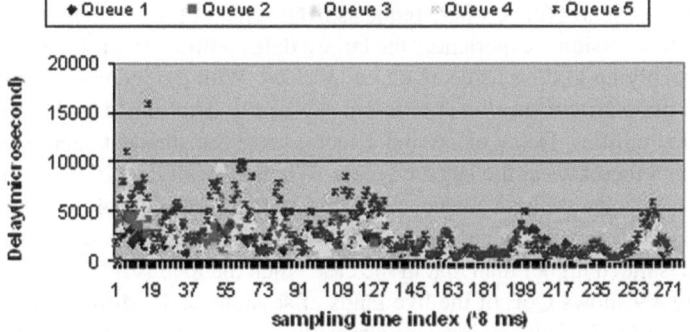

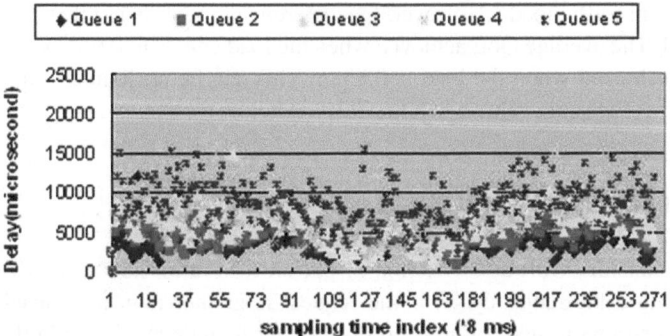

Fig. 4.8 Delay vs. traffic load

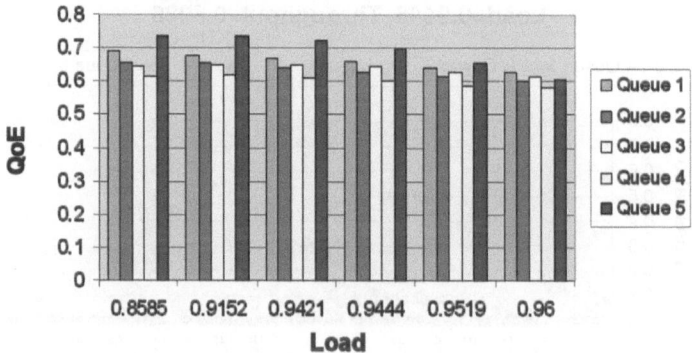

Fig. 4.9 QoE as a function of packet delay vs. traffic load

$$u^1_{i,1}(loss) = \begin{cases} 1 & loss \leq 0.01 \\ (1-loss)/0.99 & loss \in [0.01, 1] \end{cases}, \forall i$$

$$u^1_{i,2}(loss) = \begin{cases} 1 & loss \leq 0.1 \\ (1-loss)/0.9 & loss \in [0.1, 1] \end{cases}, \forall i$$

$$u^1_{i,3}(loss) = \begin{cases} 1 & loss \leq 0.2 \\ (1-loss)/0.8 & loss \in [0.2, 1] \end{cases}, \forall i$$

$$u^1_{i,4}(loss) = \begin{cases} 1 & loss \leq 0.3 \\ (1-loss)/0.7 & loss \in [0.3, 1] \end{cases}, \forall i$$

$$u^1_{i,5}(loss) = \begin{cases} 1 & loss \leq 0.4 \\ (1-loss)/0.6 & loss \in [0.4, 1] \end{cases}, \forall i$$

Figure 4.10 shows the sampled packet loss ratio of five kinds of sessions, each of which assumes one of the above five different QoE functions. The sampling is taken every 8 ms. Simulations show that packet loss happens during some of the cycles when the network load equals to 0.9446, whereas packet loss happens during most of the cycles when the network load is 1.3099. It is also shown that five kinds of sessions experience different packet loss ratios during heavily-loaded cycles. From application functions, we know that QoE of the five sessions equal to the highest value of 1 when the packet loss ratios of session 1, 2, 3, 4, and 5 are below 0.01, 0.1, 0.2, 0.3, and 0.4, respectively. From Fig. 4.10, we can see that almost all points comply with this rule. On the other hand, when the network is heavily loaded and the highest QoE score cannot be guaranteed for sessions, the packet loss ratio of session 1, 2, 3, 4, and 5 will be increased to be higher than 0.01, 0.1, 0.2, 0.3, and 0.4, respectively. For fairness, this increase should enable the five sessions achieve the same QoE. This is also substantiated in the simulation results. Therefore, in terms of the packet loss ratio, our algorithm can guarantee fairness among the five sessions.

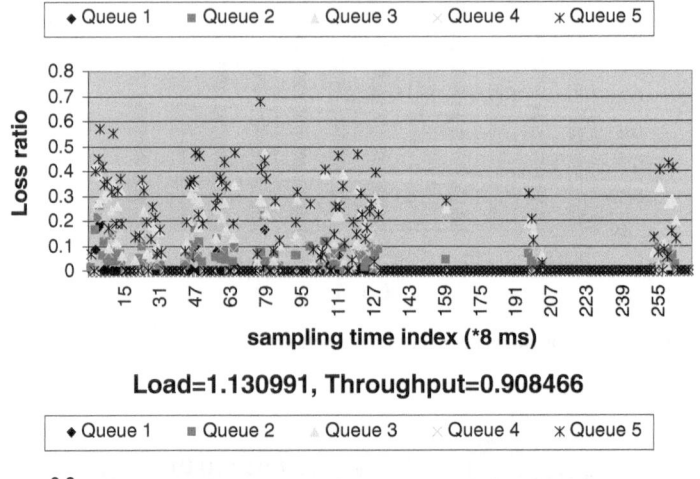

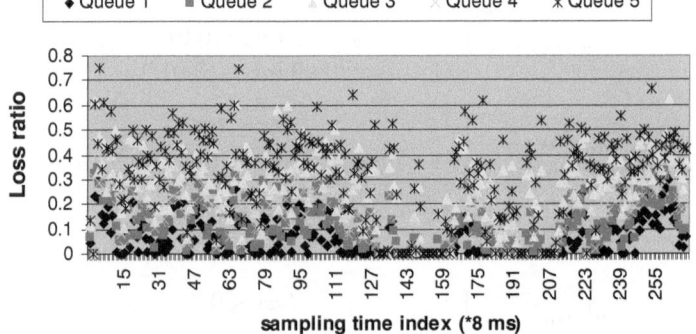

Fig. 4.10 Sampled packet loss ratio

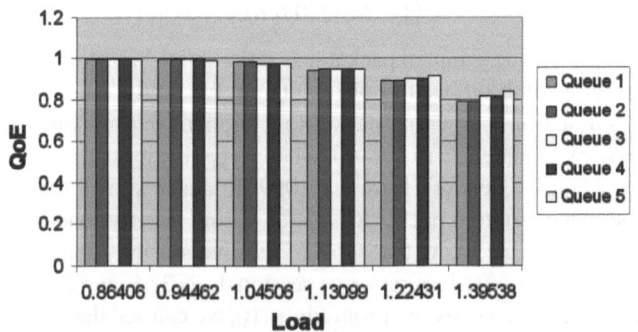

Fig. 4.11 QoE as a function of packet loss ratio vs. traffic load

Figure 4.11 shows QoE scores of the five kinds of sessions under different loads. It shows that, under a particular traffic load, all sessions are almost the same with slight differences. QoE of sessions decreases with the increase of traffic load because of the increased packet loss ratio. When the traffic load is less than 1, QoE scores of all sessions approach the maximum value of 1. When traffic load is greater than 1, QoE of sessions decrease slightly at nearly equal degrees.

4.4 Summary

In this chapter, we have discussed the MAC and resource allocation of EPON systems. The MultiPoint Control Protocol (MPCP) is standardized as the MAC layer protocol for EPON by IEEE802.3ah. MPCP describes messages, state machines, and timers to control the channel access of ONUs. MPCP performs many functions including auto-discovery, ONU registration, ranging, bandwidth polling, and bandwidth assignment. These functions are performed using a set of 64-byte control messages: GATE, REPORT, REGISTER_REQUEST, REGISTER, and REGISTER_ACK.

The EPON standard does not specify traffic scheduling algorithms for EPON and 10G-EPON. The bandwidth arbitration among different ONUs is referred to as inter-ONU traffic scheduling, while the bandwidth allocation among queues in the same ONU is referred to as intra-ONU traffic scheduling. Many schemes have been proposed to address both the intra-ONU scheduling and inter-ONU scheduling problems. The last part of this chapter details a resource allocation algorithm which maximizes user quality of experience (QoE).

Chapter 5
Media Access Control and Resource Allocation in WDM PON

By taking advantage of the huge bandwidth provision of wavelength channels, wavelength division multiplexing (WDM) passive optical network (PON) has become a promising future-proof access network technology to meet the rapidly increasing traffic demands resulted from the popularization of Internet and sprouting of bandwidth-demanding applications [7, 126]. In WDM PON, a number of wavelengths are used to provision bandwidth to ONUs in both upstream and downstream, rather than sharing a single wavelength in each stream in TDM PON. Since ONUs may share the usage of a number of wavelengths, a MAC layer control protocol is needed to coordinate the traffic transmission such that the collision between traffic from different ONUs can be avoided.

5.1 WDM PON Architecture

Generally, the available WDM PON architectures can be divided into two classes: laser-sharing-based networks and wavelength-sharing-based networks.

Laser-sharing-based networks refer to those in which ONUs share the usage of a set of lasers. One typical example is SUCCESS [3] (Fig. 5.1), which equips the central office with tunable lasers and an arrayed waveguide grating (AWG), and ONUs with WDM filters and burst-mode receivers. The wavelengths from the OLT can reach the ONUs through different PONs. All these tunable lasers are shared by ONUs, and they communicate with an ONU by tuning to the particular wavelength accommodated by that ONU. In the laser-sharing-based approach, although the ONUs have their respective dedicated wavelengths, they cannot transmit data traffic independently as they need to share the usage of lasers in the TDM fashion. From the perspective of the MAC layer, an important problem with this network is to dynamically assign these tunable lasers to ONUs to accommodate their respective traffic demands. The architecture discussed in [9] constitutes another example. In these networks, the OLT provides seed light to ONUs for the upstream transmissions of ONUs.

N. Ansari and J. Zhang, *Media Access Control and Resource Allocation: For Next Generation Passive Optical Networks*, SpringerBriefs in Applied Sciences and Technology, DOI 10.1007/978-1-4614-3939-4_5, © The Authors 2013

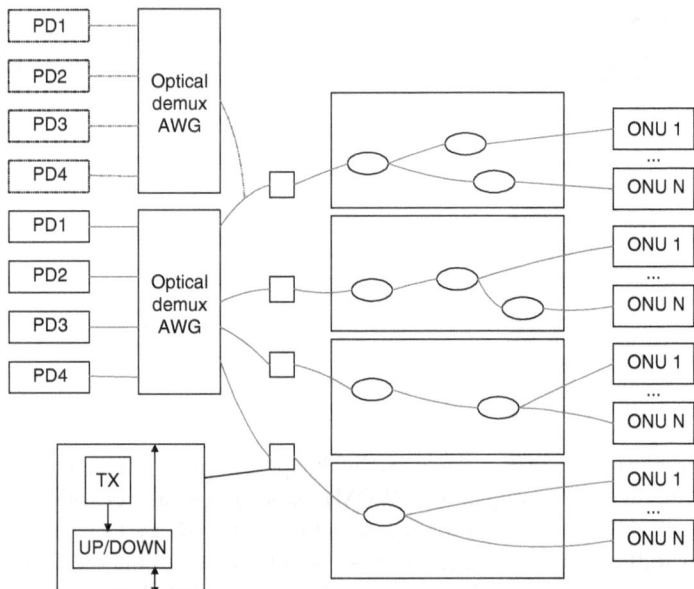

Fig. 5.1 SUCCESS architecture

In wavelength-sharing-based networks, wavelengths are shared by multiple ONUs. One example is the one proposed in [83] where signals from ONUs are first time division multiplexed onto a wavelength, and then wavelength division multiplexed onto the same fiber. In another example [114], each ONU supports two wavelengths, among which one wavelength is dedicated for this ONU and the other one is shared by other ONUs. Some, among many, have been considered as candidate architectures for next generation access stage 1 [135]. They possess the common characteristic of wavelength sharing. In these networks, ONUs are usually equipped with lasers to generate signals for their own upstream traffic transmission. They possess the common characteristic of wavelength sharing. Usually, ONUs are equipped with tunable lasers or a set of fixed-tuned lasers. Each ONU can access some wavelengths depending on the wavelengths supported by its lasers.

For wavelength-sharing-based networks, depending on the wavelength generation capability, there are three major classes of lasers available for use—namely, multiwavelength lasers, wavelength-specified lasers, and wavelength-tunable lasers [62]. A multiwavelength laser is able to generate multiple WDM wavelengths simultaneously, including multifrequency laser, gain-coupled distributed feedback laser diode (DFB LD) array, and chirped-pulse WDM. Multiwavelength lasers are usually used at the optical line terminal (OLT) to generate downstream traffic or seed optical network units (ONUs) with optical signals for their upstream data transmission [84, 121]. Instead of generating multiple wavelengths, a wavelength-specified laser, e.g., the common DFB, can only emit one specific wavelength. However, with wavelength-specified lasers, no statistical gain can be exploited

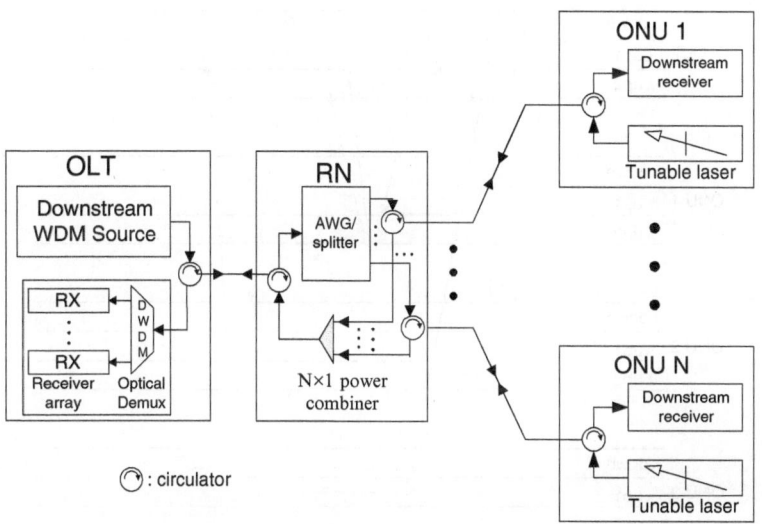

Fig. 5.2 Wavelength-sharing based architecture (Fig. 1 in [125])

among ONUs that can support different wavelengths. Wavelength-tunable lasers are able to generate multiple wavelengths, but only one wavelength at a time. Tunable lasers possess advantages of facilitating the statistical multiplexing of traffic from all ONUs and enable the color-free property of ONUs. Figure 5.2 shows an example of the WDM PON with tunable lasers.

The network architecture and the wavelength supports of different ONUs affect the Media Access Control and resource allocation of the PON system. More specifically, the issues of accessing and dynamically assigning network resources to ONUs for their data transmissions have to be addressed [17, 77]. Laser-sharing-based networks require dynamic and realtime laser assignment schemes, while wavelength-sharing-based networks need dynamic wavelength assignment algorithms. Resource allocation problems in both networks can be modeled as multiprocessor scheduling problems by considering lasers and wavelengths as processors in the two respective cases [125, 126, 129, 133]. The following addresses the wavelength assignment problem for the latter class of networks. Similar strategies may be applied to solve the laser assignment problem for the former class of networks.

5.2 Media Access Control

In WDM PONs, traffic from multiple ONUs is multiplexed in the TDM fashion onto wavelength channels. Owing to the shared nature of the wavelength channel, WDM PONs require proper media access control (MAC) protocols to coordinate

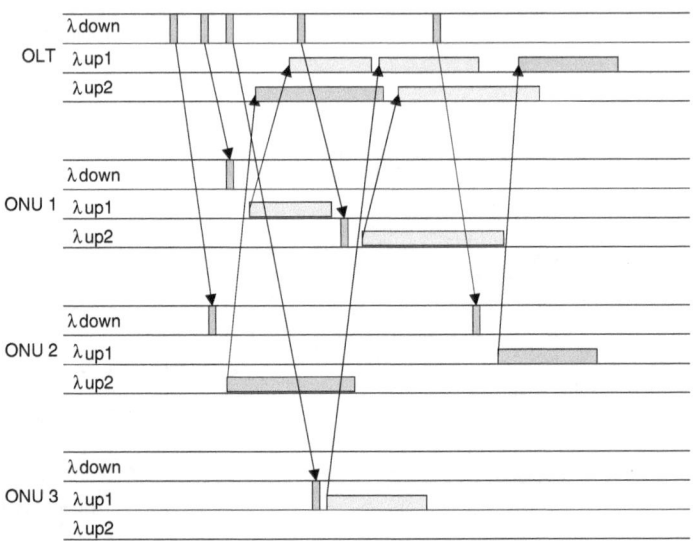

Fig. 5.3 WDM IPACT-ST

communications between the OLT and ONUs so that the collision of data transmissions from more than one ONU can be avoided.

For the downstream transmission in WDM PONs, the downstream incoming packets are queued in buffers at the OLT upon arrivals. Then, the OLT determines the downstream bandwidth allocated to ONUs, and sends the downstream packets out to ONUs. Different from the downstream transmission, the upstream transmission in hybrid WDM/TDM PONs needs a proper MAC protocol to avoid data collision among ONUs. For backward compatibility, the MAC layer control protocol of hybrid WDM/TDM PONs inherits some characteristics from those of EPON and GPON, two major flavors of the existing TDM PONs. The data transmission processes of the two PONs are similar and can be generalized as follows: ONUs report their queue lengths and send their data packets to the OLT using time slots allocated by the OLT; the OLT collects queue requests, makes bandwidth allocation decisions, and then notifies ONUs when and on which channel they can transmit packets. Such a request-grant based transmission mechanism is highly likely to be adopted in WDM PONs to be backward compatible with both EPON and GPON [17,60,76]. Following the assumption of a request-grant based MAC control mechanism, the OLT gathers most of the intelligence and control of the network, and its functions determine the performance of the network.

Kwong et al. [60] proposed a multiple upstream wavelength version of IPACT [57], referred to as WDM IPACT-ST (see Fig. 5.3). This algorithm assumes each ONU supports all wavelengths. It keeps track of the available time on each upstream wavelength. Upon receiving a REPORT from an ONU, the OLT schedules the ONU's next transmission grant on the next available wavelength. Considering the case that an ONU may not support all available wavelengths, McGarry et al. [76] proposed to select the next available supported wavelength for scheduling.

McGarry et al. [77] further introduced the concept of the scheduling framework and scheduling policy to address the issues on when and how the OLT performs DBA, respectively. Three scheduling frameworks were defined, i.e., on-line scheduling, off-line scheduling, and just-in-time scheduling. On-line scheduling refers to the operation that the OLT determines bandwidth allocated to an ONU immediately after receiving this ONU's request; off-line scheduling refers to the operation that the OLT performs DBA after receiving queue requests from all ONUs. Both on-line scheduling and off-line scheduling have their advantages and disadvantages. On-line scheduling enables ONUs receive immediate grants. However, the bandwidth allocation decision is made based on only one ONU's request. This may result in unfairness for other ONUs with upcoming requests. Off-line scheduling achieves better fairness by making decisions based on the requests of all ONUs. However, it incurs delays for ONUs to receive grants, and under-utilizes the gap between the time that the OLT sends out the grant and the time that the OLT receives the report from the first ONU.

To overcome the near-sight problem of on-line scheduling and the under-utilization problem of off-line scheduling, McGarry et al. [77] proposed just-in-time scheduling, where the OLT postpones the decision making time until one channel is about to become idle. The decision making time in just-in-time scheduling is later than that in on-line scheduling, and is earlier than that in off-line scheduling. These three scheduling frameworks show similar advantages and disadvantages when they are applied in downstream scheduling.

5.2.1 Dynamic Bandwidth Allocation

The scheduling policy addresses the problem of deciding how to allocate bandwidth to ONUs. It involves two problems: wavelength assignment and time allocation. For wavelength assignment, the earliest-channel-available-first rule was proposed to be employed with the assumption that each ONU can support all wavelengths[17, 60]. To make the algorithm applicable to the case that ONUs may support only a subset of the wavelengths, McGarry et al. [76] modified the earliest-channel-available-first rule into the next-available-supported-channel-first. McGarry et al. [77] converted the wavelength assignment problem into a matching problem between wavelengths and ONUs, and proposed Weighted Bipartite Matching to solve the matching problem. McGarry et al. [78] modeled the problem as a multiprocessor scheduling problem and proposed to use the longest processing time (LPT) first rule to address the makespan minimization problem for the case that ONUs can access all the wavelengths. When ONUs can access a limited set of wavelengths, they are scheduled according to the least flexible job (LFJ) first rule. Meng et al. [80] studied the joint grant scheduling and wavelength assignment problem. They formulated it into a mixed integer linear programming (MILP) problem, and employed tabu search to obtain the optimal solution. For the time allocation problem, the time allocated to ONUs usually equals to its corresponding request when the on-line scheduling framework is adopted. In off-line scheduling, Dhaini et al. [17] proposed three time allocation algorithms, whereby low-load ONUs can always have their requests satisfied and high-load ONUs share the excess bandwidth by using different methods.

5.3   Modeling and Problem Formulation

The general bandwidth allocation problem in WDM PONs in a single DBA cycle can be described as:

> Given the wavelength access capability of ONUs, the time that each ONU takes to switch wavelength channels, requests from n ONUs, the available time of m wavelength channels, and the wavelength initially accessed by each ONU, construct a schedule such that all requests can be accommodated.

The bandwidth allocation problem can be mapped into a multiprocessor scheduling problem [37], with ONU requests as jobs and wavelength channels as machines. Jobs and machines possess their respective unique characteristics.

5.3.1   Wavelengths → Machines

Wavelength channels are modeled as parallel machines. Considering the case that the data rate on each wavelength may not be the same, different machines may have different processing time for the same request from an ONU. Note that these wavelength channels may not be simultaneously available.

5.3.2   ONU Requests → Jobs

There are two options to model jobs. The first one is to model each queue request of an ONU as an individual job. However, owing to the laser on/off time, some guard time is needed between scheduling of jobs from different ONUs. To save the guard time, jobs from the same ONU is desired to be scheduled consecutively, and thus can be grouped together as a single job for simplicity. The second option is to regard the total requests of an ONU as a single job. Then, jobs possess two main properties. A job can be divided into subjobs, corresponding to requests of queues in the ONU, and each subjob can be further divided into subjobs, corresponding to requests of packets in the queue. In GPON with the allowance of packet fragmentation, scheduling of a packet can even be divided into scheduling of its partial packets, while in EPON without packet fragmentation, scheduling of a packet cannot be further divided. Therefore, jobs are preemptable in hybrid WDM/TDM GPON, and preemptable to certain degree in hybrid WDM/TDM EPON. In the case that ONUs are equipped with tunable lasers and tunable lasers require a non-negligible amount of time to switch wavelengths, certain time gap is needed between the scheduling of jobs from the same ONU on different wavelength channels.

In addition, in multiprocessor scheduling, preemption enables jobs to be scheduled more flexibly, thus yielding better system performances as compared to non-preemption. However, when preemption is allowed, jobs may be divided and scheduled in non-continuous time periods; this incurs some extra time gap for laser on/off. It is not easy to tell whether the extra cost introduced by the guard time outweighs the extra performance improvement introduced by flexibility.

5.3.3   Scheduling Objective

In assigning resources to ONUs, typically, small delay of ONU packets, fairness among ONUs, and load balancing on wavelengths are desired to be achieved. Minimizing the average delay of ONU packets is equivalent to minimizing the average job finishing time in multiprocessing scheduling. Achieving load balancing on wavelength channels can be mapped into the problem of minimizing the latest job completion time on all wavelengths channels. Fairness among ONUs are desired when the network resources are not enough to accommodate all ONUs' requests that some packets have to be dropped or delayed. In this case, ONUs should have an equal proportional amount of packets to be delayed or dropped for fairness concern. In addition, in terms of the just-in-time scheduling framework, wavelength channels may need to become idle simultaneously to the best for the following reasons. The OLT makes bandwidth allocation decisions before any of the wavelengths becomes idle. If all wavelengths become idle simultaneously, the scheduler can collect the requests from most of the ONUs, and can thus make a fair decision. If one wavelength turns idle much earlier than the others, few requests arrive at the scheduler before the decision making time. In the worst case, just-in-time scheduling may be degraded into on-line scheduling, which makes the decision for one ONU request only. This will result in unfairness and increase the frequency of calculating bandwidth allocation.

In the following, we investigate two resource allocation problems [130, 133] and their mappings into multiprocessor scheduling problems in WDM PON.

5.4   Problem 1: Non-preemptive Scheduling for ONUs Supporting Different Sets of Wavelengths

Graham et al. [26] proposed a three-field $\alpha|\beta|\gamma$ classification scheme for multiprocessor scheduling problems, in which the α field describes the machine environment, the β field describes the job characteristics, and the γ field is the objective function. Following the well known three-field $\alpha|\beta|\gamma$ classification scheme, the scheduling problem under this case is equivalent to the $p||C_{max}$ multiprocessor scheduling problem, which consists in finding an optimal assignment of tasks/jobs to be processed by the set of processors/machines, p, such that the makespan, C_{max}, of the assignment is minimized. First, consider the case that the guard time equals to zero. Then, the time requirement of ONU i equals to the time duration for its data transmission, denoted as $\mathbf{r}_i$. Denote C^* and C^H as the minimum makespan and the makespan achieved by heuristic algorithm H, respectively. An algorithm is referred to as a ρ approximation algorithm if its makespan is no greater than ρ times that of the optimal makespan for all instances, i.e., $C^H \leq \rho \cdot C^*$. Many heuristic algorithms have been proposed for the $p||C_{max}$ problem. For example, list scheduling constructs a list of requests, and schedules these requests in order by using the earliest available wavelength channel; Longest Processing Time (LPT) list scheduling modifies list scheduling by ordering requests with the descending order of their sizes first. These two algorithms were shown to achieve an approximation ratio of $2 - 1/m$ and $4/3 - 1/3m$, respectively. Coffman et al. [16] proposed another algorithm referred to as the multifit algorithm. Although the multifit algorithm cannot be guaranteed to obtain a better performance than LPT for all instances, it was shown that multifit achieves an approximation ratio of $72/61$, which is smaller than that of LPT list scheduling. Hence, we suggest to use the multifit algorithm in the wavelength scheduling of hybrid WDM/TDM PON.

Considering the guard time between the scheduling of ONUs, we change the time requirement of ONU i from $\mathbf{r}_i$ to $\tilde{\mathbf{r}}_i = \mathbf{r}_i + \mathfrak{g}$, where $\mathfrak{g}$ is the guard time between the scheduling of two ONUs, and it includes the time to turn laser on/off, to facilitate automatic gain control (AGC), to enable clock and data recovery (CDR), and to accommodate the MAC layer overhead. The problem is still equivalent to the $p||C_{max}$ problem, which can be addressed by the same algorithm as that in the case that the guard time equals to zero.

Similar to the scenario that ONUs have full wavelength access capability, we consider $\mathbf{r}$ as ONU requests in the case that the guard time equals to zero, and regard $\tilde{\mathbf{r}} = \mathbf{r} + \mathfrak{g}$ as ONU requests in the case that the guard time does not equal to zero. When ONUs can only access a limited set of wavelengths, the scheduling is equivalent to the $p|M_j|C_{max}$ multiprocessor scheduling problem, where M_j describes the set of eligible machines for job j. The $p|M_j|C_{max}$ problem is more general than the $p||C_{max}$ problem, and is hence NP-hard.

When the supported wavelengths of ONUs (eligible machines) are nested, the problem is simpler than that with arbitrary wavelength supportability. Formerly, Centeno and Armacost [13] developed a heuristic algorithm for the problem that integrates the least flexible job first rule (LFJ) and the least flexible machine first rule (LFM). Here, the LFJ rule selects the job that can be processed with the smallest number of machine types first, and the LFM rule assigns the job to the most restricted machine. Pinedo [87] stated that the LFJ rule was optimal for $Pm|p_j = 1, M_j|C_{max}$ when the M_j sets are nested, where $p_j = 1$ denotes that all jobs are of the unit processing time.

For the general case with arbitrary eligible machine constraints, Centeno and Armacost [14] developed some heuristic algorithms for the $P_m|r_j,M_j|C_{max}$ problem, where r_j is the release time of job j. They showed that the rule used for job selection affects the performance of heuristic algorithms, and that the LPT rule is superior to the LFJ rule when the machine eligibility sets are not nested. Potts [88] developed a two-approximation algorithm by using a relaxed linear programming with rounding technique. The rounding process takes 2^m steps. Lenstra et al. [66] modified the rounding technique to eliminate the exponential computation. They also showed that no polynomial algorithm can achieve a worst-case ratio less than $3/2$ unless $P = NP$. Shchepin et al. [102] further improved the rounding process and developed a $2 - 1/m$ approximation algorithm. To the best of our knowledge, this is the best approximation algorithm for this problem.

When $m = 2$, C^H is as large as 1.5 times that of C^* in the worst case; C^H approaches two times that of C^* in the worst case with the increase of the number of wavelengths. Motivated to obtain a smaller makespan, we next investigate the preemption version of the problem.

5.5   Problem 2: Preemptive Scheduling for Lasers in ONUs Requiring Non-zero Tuning Time

When lasers require non-zero tuning time, the scheduling problem is equivalent to the preemptive multiprocessor scheduling problem with the objective of minimizing the makespan subject to the constraints that machines are non-simultaneously available and jobs take non-negligible time to switch machines.

Denote τ as the laser tuning time, and C_m^{-1} as the latest job completion time on wavelength m in the last cycle. Denote $\alpha_{w,i}$ as the earliest time that wavelength w can be allocated to the request from ONU i. We also denote $\alpha_{w,i}$ for request from ONU i whose wavelength was tuned to wavelength w in the last cycle as a_w^l. When $\tau = 0$ and all wavelengths channels are available at the same time, i.e., $a_w^l = a_{w'}^l, \forall w \neq w'$, the problem is equivalent to the $p|pmtn|C_{max}$ multiprocessor scheduling problem [87], which can be easily solved. When $\tau = 0$ and wavelength channels are not simultaneously available, i.e., $\exists w \neq w', a_w^l \neq a_{w'}^l$, the problem can be solved by slightly modifying the algorithm for the $p|pmtn|C_{max}$ problem.

When the laser tuning time $\tau = +\infty$, the request from ONU i can only be scheduled on the original wavelength λ_i^{-1} tuned by ONU i. The latest job completion time on wavelength w equals to $a_w^l + \sum_{\{i|\lambda_i^{-1}=w\}} r_i$. Among all wavelengths, the latest job completion time equals to $\max_{w=1}^m \left(a_w^l + \sum_{\{i|\lambda_i^{-1}=w\}} r_i\right)$.

When the laser tuning time τ is an arbitrary value, the preemptive scheduling problem was investigated in [130] and was shown to be NP-hard. For completeness, we provide the proof below.

Theorem. *When the laser tuning time τ is arbitrary, the preemptive scheduling problem with the objective of minimizing the latest request completion time is NP-hard.*

Proof. Consider the following downstream traffic scheduling problem with $C_w^{-1} = \begin{cases} t + \tau & \text{if } w = 1 \\ t & \text{otherwise} \end{cases}$, $\lambda_i^{-1} = 1 \ \forall i$, and $r_i \in [\ell - t - 2\tau, \ell - t - \tau], \forall i$. Then, after checking all i and w, $\alpha_{i,w} = t + \tau, \forall i, w$. Since $\alpha_{i,w} + r_i \leq \ell$ and $\alpha_{i,w} + r_i + \tau \geq \ell$, any request can be scheduled on any wavelength, but cannot be divided into parts and scheduled on multiple wavelengths. The time duration which can be allocated on any wavelength equals to $(\ell - t - \tau)$. The problem of determining whether all requests can be scheduled before ℓ is equivalent to the problem of deciding whether all these given requests can be divided into m groups, in which the sum of requests in each group is no greater than $\ell - t - \tau$. The latter problem is equivalent to the bin packing problem, which is NP-hard [21]. Hence, the preemptive scheduling problem with the objective of minimizing the latest request completion time is NP-hard when the laser tuning time τ is arbitrary. □

Because of the NP-hard property, several heuristic algorithms [130] have been proposed to solve the problem. For illustrative purposes, we shall present a preemptive scheduling algorithm, referred to as *naive preemptive scheduling*, which is based on the schedules constructed for $\tau = 0$ and $\tau = +\infty$, respectively.

The main idea of naive preemptive scheduling is to first construct a schedule $\mathfrak{S}^p(\mathbf{r}, 0)$ assuming that the laser tuning time is zero, and a schedule assuming that the laser tuning time is $+\infty$. Then, naive preemptive scheduling adjusts the schedule to give all lasers enough time to switch wavelengths. If the schedule length is less than the schedule length with zero tuning time, the adjusted schedule based on the schedule with zero tuning time is considered as the final schedule; otherwise, the schedule with $+\infty$ tuning time is considered as the final schedule. Algorithm 5 (same as Algorithm 1 in [130]) details the proposed naive preemptive scheduling.

The part between Lines 2 and 10 in Algorithm 5 is to construct a schedule assuming that the laser tuning time is zero. For the scheduling algorithm with zero laser tuning time, the main idea is to try different number ℓ and decide whether there exists a schedule whose latest job completion time is ℓ. Finally, the schedule with the minimum latest job completion time can be obtained. The decision problem can be formulated as: given the latest job completion time ℓ, can all jobs be scheduled before ℓ? In computational complexity theory, the decision problem is not easier than the original optimization problem.

For a given ℓ, requests are first sorted in the descending order of their sizes, and wavelengths are sorted in the ascending order of their available time as described in Lines 3 and 4. The sorting is to make sure that large requests receive enough allocations of nonoverlapping time durations. Then, the time resource on a wavelength is assigned to requests one by one from the back of the time span until the time on that wavelength is used up. If the remaining time on a wavelength

---

Algorithm 5 Naive preemptive scheduling

1: $\ell = (\sum_i r_i + \sum_w a_w)/m$
2: **while** The smallest latest request completion time ℓ has not been found **do**
3:      Index ONU requests such that $r_1 \geq r_2 \geq \ldots \geq r_n$
4:      Index wavelengths such that $a_1^l \leq a_2^l \leq \ldots \leq a_m^l$
5:      Select an ONU request and a wavelength
6:      Schedule the request on the back of the wavelength. If the remaining time on a wavelength is not enough for the request, schedule the remaining unscheduled part of the request to another wavelength.
7:      **if** Not all requests can be scheduled before ℓ **then**
8:          Find a proper ℓ
9:      **end if**
10: **end while**
11: Postpone the scheduling of all requests on a wavelength by τ
12: Postpone the scheduling of the last request on a wavelength by τ
13: If the length of the constructed schedule is longer than the schedule with $+\infty$ tuning time, the schedule with $+\infty$ tuning time is considered as the final schedule.

---

is not enough to satisfy a request, the unscheduled part will be moved to the next wavelength as described in Line 6.

To generate a feasible schedule which gives lasers sufficient time to switch wavelengths, we perform some further adjustments. The first step is to postpone the scheduling of all requests by τ as described in Line 11. The second step is to postpone the scheduling of the last request on each wavelength by τ as described in Line 12.

We then prove that the schedule produced by naive preemptive scheduling is a feasible schedule under the condition that the laser tuning time equals to τ.

Property The schedule produced by Algorithm 5 is a feasible schedule for the case that the laser tuning time equals to τ.

Proof. Since all requests are postponed by time τ, the corresponding laser for request i is idle during $[a_w^l, a_w^l + \tau]$, and laser i is given enough time to schedule the first request. Besides, in the schedule $\mathfrak{S}^p(\mathbf{r}, 0)$, only requests scheduled at the beginning or the end of the time span of a wavelength may be preempted. In Line 12, the last scheduled request on each wavelength is postponed by τ, and hence lasers are given sufficient time to schedule the last scheduled request. $\square$

Therefore, in the schedule produced by Algorithm 5, all requests have been scheduled, and lasers are all given enough time to switch wavelengths.

Computational Complexity: The complexities of the two ordering processes are $O(n\log(n))$ and $O(m\log(m))$, respectively. The complexity of the "for" loop in Algorithm 5 is $O(n)$. Lines 11, 12, and 13 are all of complexity of $O(n)$. Hence, the total complexity of Algorithm 5 is $O(n\log(n) + m\log(m))$.

The example with 8 ONUs and 3 wavelengths as shown in Fig. 5.4 illustrates Algorithm 5. Figure 5.4a shows the constructed schedule assuming that the laser tuning time equals to zero. Request 1 is allocated with the time duration [6, 14] on

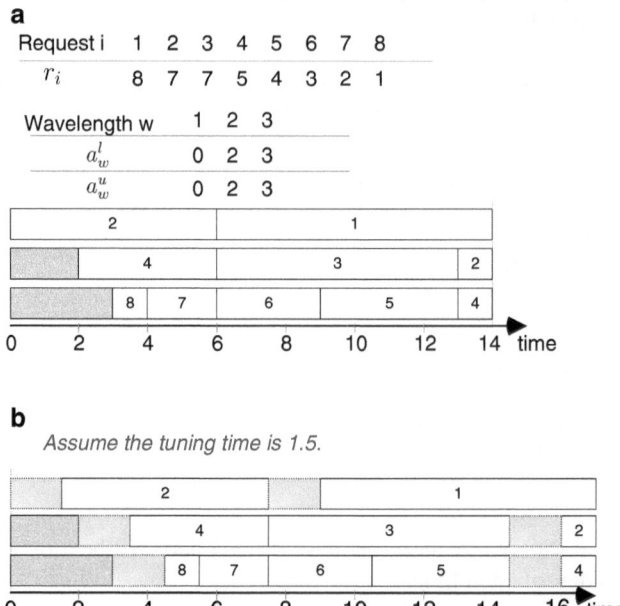

Fig. 5.4 One example of naive preemptive scheduling when $\tau = 1.5$. (Fig. 1 in [130]). (**a**) The preemptive schedule when $\tau = 0$. (**b**) The preemptive schedule when $\tau = 1.5$

wavelength 1. The remaining time duration on wavelength 1 is not enough to satisfy request 2. Part of request 2 is scheduled in time duration [0, 6] on wavelength 1 and the other part is scheduled in time duration [13, 14] on wavelength 2. Similarly, part of request 4 is scheduled on wavelength 2, and the other part is scheduled on wavelength 3. All the requests can be scheduled before time 14. Figure 5.4b shows the final schedule after adjustment. When $\tau = 1.5$, the latest job completion time is increased from 14 to 17.

5.6   Summary

In WDM PON, a number of wavelengths are used to provision bandwidth to ONUs in both upstream and downstream, rather than sharing a single wavelength in each stream in TDM PON. Although a variety of WDM architectures have been proposed, WDM PON still needs to overcome some hurdles before being extensively deployed. On the other hand, different MAC schemes have been proposed for WDM PONs to tackle the MAC layer challenge such that the wavelength channels can be accessed by ONUs in an efficient way. Extending the TDM PON MAC protocol for WDM PON systems is one of the major methods. Allocating network resources including both wavelength channels and time slots is another challenge issue in WDM PON. Generally, the joint wavelength assignment and time allocation problem can be modeled as a multiprocessor scheduling problem. We have respectively analyzed resource allocation problems for two WDM PONs in this chapter. Readers are referred to [130, 133] for other heuristic scheduling algorithms for hybrid WDM/TDM PONs.

Chapter 6
OFDM PON

Orthogonal frequency division multiple access PON (OFDMA PON) employs orthogonal frequency division multiplexing (OFDM) as the data modulation scheme to increase the provisioning data rate. Qian et al. [89] demonstrated the transmission of 108 Gb/s downstream data rate over a type B fiber with a split ratio of 32. Figure 6.1 describes one typical OFDMA PON architecture. OFDMA PON uses OFDMA as an access scheme [90, 111], which divides the upstream/downstream bandwidth in baseband into multiple subcarriers with orthogonal frequencies. These subcarriers are dynamically allocated to different ONUs based on their real-time incoming traffic information. As compared to TDM PON and WDM PON, OFDMA PON enjoys numerous advantages such as high speed transmission, fine granularity of bandwidth allocation, and color-free ONUs [137].

6.1 OFDM PON Architecture

OFDMA PON combines both OFDM and TDMA. As compared to TDM PONs, OFDMA PON is potentially able to achieve a much higher bandwidth provisioning. The dynamic subcarrier allocation property enables the bandwidth sharing among different ONUs and applications, and therefore further enhances the network performance. As compared to WDM PONs, OFDMA PON has much finer granularity of bandwidth allocation. OFDMA PON is potentially able to increase the system reach and transmission rates without increasing the required cost/complexity of optoelectronic components.

6.2 Media Access Control

The upstream/downstream data traffic is transmitted over one wavelength channel, which is further divided into OFDM subcarriers. Each OFDMA subcarriercan be

N. Ansari and J. Zhang, *Media Access Control and Resource Allocation: For Next Generation Passive Optical Networks*, SpringerBriefs in Applied Sciences and Technology, DOI 10.1007/978-1-4614-3939-4_6, © The Authors 2013

Fig. 6.1 OFDMA PON
architecture

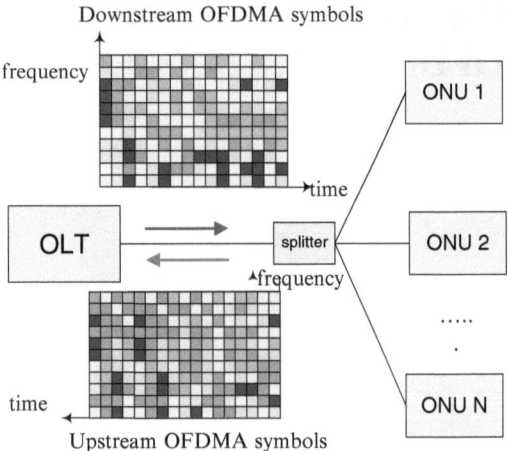

allocated to different ONUs in different time slots. To avoid collision in accessing
upstream OFDMA subcarriers, proper control schemes are desired to coordinate
data transmissions of ONUs.

6.2.1 TDMA-Based Media Access Control

One natural solution of the MAC protocol is to adapt current MAC protocols of
TDM PONs to OFDMA PON. Basically, TDM PONs including EPON and GPON
employ the following upstream bandwidth control scheme: ONUs report their queue
length information to the OLT using the time slot specified by the OLT; the OLT
allocates time slots in a frame/cycle to ONUs and notifies ONUs on its decisions;
ONUs transmit their data traffic over time slots allocated by the OLT.

As shown in Fig. 6.2, with the above scheme, all subcarriers are shared among all
ONUs. Thus, statistical multiplexing gain among traffic of ONUs can be efficiently
exploited. However, it requires a rather complex inter-ONU bandwidth arbitration
scheme. Moreover, similar to legacy TDM PONs such as EPON and GPON,
ONUs are required to be synchronized with the OLT before fulfilling their data
transmission and receiving.

6.2.2 FDMA-Based Media Access Control

Another solution is to divide all OFDMA subcarriers into multiple nonoverlapping
sets, each of which is fixedly allocated to an ONU as illustrated in Fig. 6.3. Since
no sharing of OFDMA subcarriers exists among different ONUs, ONUs can send

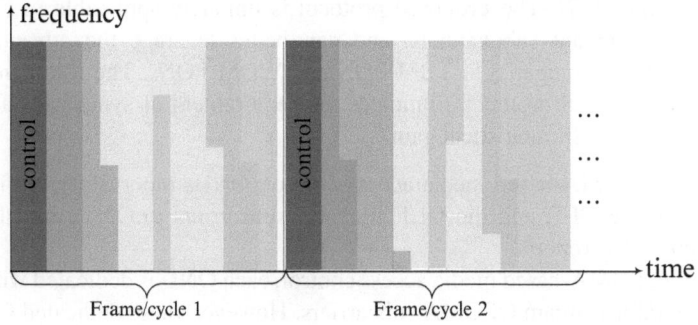

Fig. 6.2 TDMA based media access control

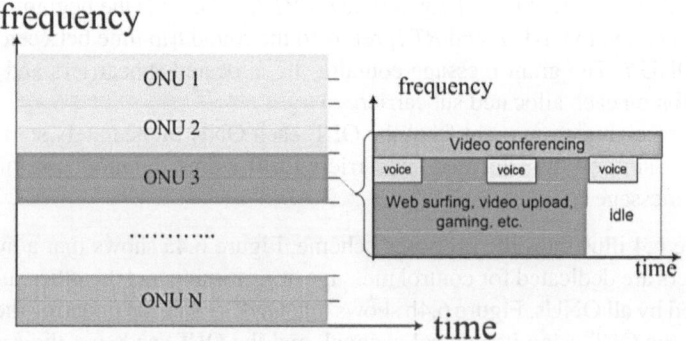

Fig. 6.3 FDMA based media access control

their traffic over the dedicated subcarriers any time they want without collision. The communication between the OLT and an ONU can be actually regarded as a point-to-point system. The elimination of the requirement of synchronization, media access control protocols, and sophisticated inter-ONU bandwidth arbitration algorithms simplifies the ONU structure, and thus reduces the ONU cost. Low bandwidth utilization and therefore low network performance will be resulted owing to the absence of statistical multiplexing gain. The low bandwidth utilization problem is not negligible particularly in PONs where the ONU traffic exhibits burstiness and strong self-similarity, which are characteristics of many user applications such as variable bit rate videos.

6.2.3   Hybrid TDMA and FDMA Media Access Control

The hybrid TDMA and FDMA MAC protocol can be adequately tailored for OFDMA PON by capitalizing on the abundance of OFDMA subcarriers to facilitate asynchrony of ONUs and at the same time to exploit the statistical multiplexing gain of ONU traffic [137]. The proposed protocol is uniquely applicable in OFDMA PONs with abundant subchannels, and can better leverage the advantages of OFDMA PON as compared to TDM PON and WDM PONs. The following novel media access control protocol eliminates the requirement of synchronization and also exploits the traffic statistical gain.

- Similar to TDMA-based media access control, ONUs report their traffic information to the OLT, and the OLT allocates subcarriers to ONUs based on the real-time ONU reports.
- Similar to FDMA-based media access control, each ONU is dedicated with some upstream/downstream OFDMA subcarriers. However, these dedicated OFDMA subcarriers are used for control message transmission only.
- The OLT sends out the grant message to an ONU right before the allocated time begins, i.e., the grant is sent out at time $t - RTT_i$, where t is the beginning time of the allocation to ONU i, and RTT_i refers to the round trip time between the OLT and ONU i. The grant message contains the allocated subcarriers and the time duration on each allocated subcarrier.
- Upon receiving grants sent from the OLT, each ONU immediately starts its data transmission on the allocated subcarriers for the time duration specified in the grant message.

Figure 6.4 illustrates the proposed scheme. Figure 6.4a shows that a number of subcarriers are dedicated for control message transmission and the other subcarriers are shared by all ONUs. Figure 6.4b shows that ONUs can keep updating their queue status to the OLT using its control channel, and the OLT sends out the grant to an ONU just before the allocation time begins. We assume that RTT_i ($\forall i$) is known to the OLT during the ONU registration process.

Generally, the proposed scheme exhibits three main advantages. First, with dedicated upstream control subcarriers for each ONU, an ONU can update its latest queue status to the OLT and make sure the OLT get the most recent queue information. Second, with the dedicated downstream control subcarriers, the OLT can send the grant just before the allocated time begins such that ONUs can begin transmission immediately after having received grants without synchronizing with the OLT clock. Third, all the other OFDM subcarriers except control subcarriers are shared by ONUs, thus facilitating the statistical multiplexing gain.

6.2.4   Performance Evaluation

This section investigates the packet delay and throughput performances of the above described media access control scheme. The simulation setup is as follows. Assume the PON supports 32 ONUs, and ONUs are 20 km away from the OLT. $RTT_i, \forall i$, is set as 0.2 ms. The upstream/downstream data rate is set as 10 Gb/s, and we consider 2048 OFDMA subcarriers, among which each ONU is dedicated with one subcarrier for control message transmission. Then, each ONU is allocated with 4.88 Mb/s

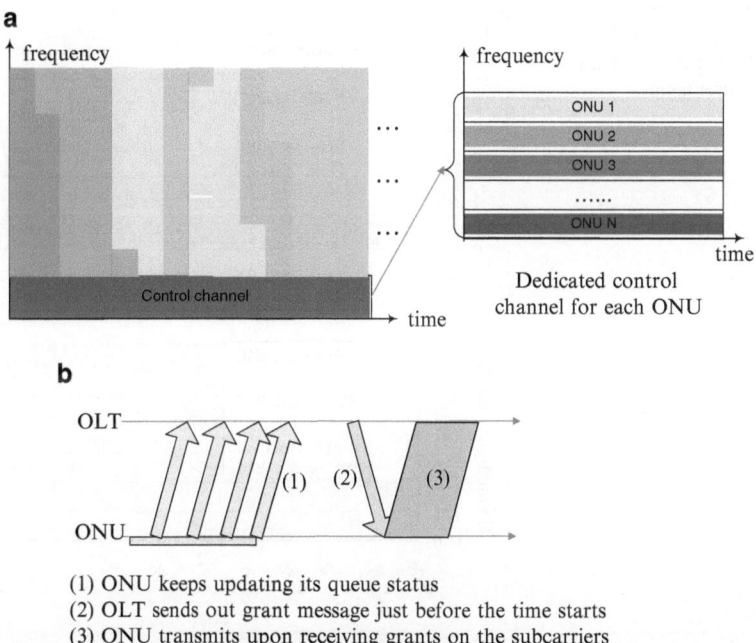

(1) ONU keeps updating its queue status
(2) OLT sends out grant message just before the time starts
(3) ONU transmits upon receiving grants on the subcarriers
    for the time duration specified in grant message

Fig. 6.4 The proposed subcarrier allocation and media access control scheme

upstream/downstream bandwidth for control traffic, and an ONU can update its queue information every 10.5 μs if the length of the report message equals to 64 bytes. For the ONU traffic, a finite-time horizon with the time duration of 8 s is chosen. We assume the traffic of an ONU arrives in bursts. The burst size obeys the Pareto distribution with the Pareto index $\alpha = 1.4$ and the mean equals to 31.25 k bytes; it takes about 25 μs to transmit 31.25 k bytes if all OFDMA subcarriers, except those dedicated for control messages, are used for the transmission. The burst inter-arrival time also obeys the Pareto distribution with $\alpha = 1.4$. The mean is varied to produce different network traffic loads.

Figure 6.5a compares the delay performance produced by the three MAC control schemes. Traffic load is defined as the ratio between the total arrival traffic over the network capacity. In FDMA-based media access control, each ONU is fixedly assigned with $2048/32 = 64$ subcarriers for their data transmission. In the other two schemes, all subcarriers except those dedicated for control messages are allocated to the same ONU at a time. Thus, FDMA-based media access control produces longer packet transmission delay as compared to the other two schemes. When the network is lightly loaded, the transmission delay dominates the overall delay, and hence FDMA-based media access control yields the largest delay among the three schemes under this traffic condition. Besides, in our proposed scheme, traffic arrival can be immediately reported to the OLT while the incoming traffic has to wait for

Fig. 6.5 Performance comparison between three control schemes: (**a**) delay, and (**b**) throughput

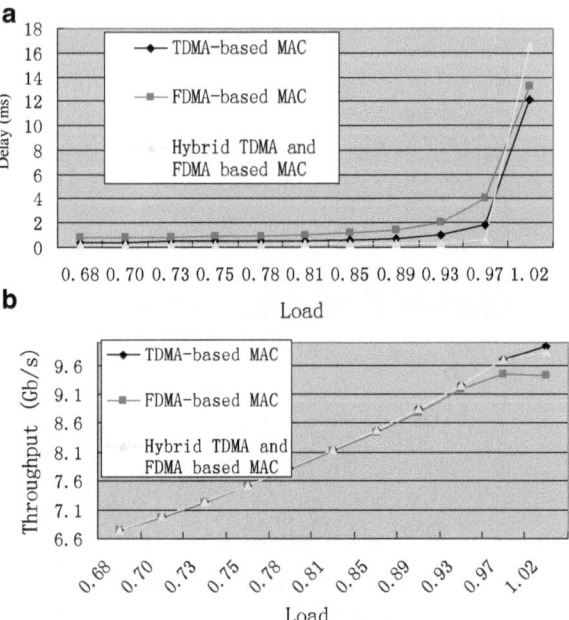

some time before being reported in FDMA-based MAC control. Thus, when the network is lightly loaded that queuing delay is negligible, our proposed scheme yields smaller delay than FDMA-based MAC control. Figure 6.5a shows that the proposed scheme produces the smallest delay when the network load is as large as 0.97. When the network is heavily loaded ($load > 0.97$ as shown in Fig. 6.5a), the proposed scheme results in the largest delay because of the large queuing delay resulted by the reduced number of subcarriers for data transmission.

Figure 6.5b compares the throughput performance of the three schemes. When $load < 0.93$, throughputs of the three schemes are similar, and are equal to the arrival traffic rates; when $load > 0.93$, throughput of FDMA-based media access control is the smallest because of the failure of exploiting the statistical multiplexing gain, and throughput of our proposed scheme is slightly smaller than that of TDMA-based media access control because 32 OFDMA subcarriers are dedicated for control message transmission.

6.3   Summary

OFDMA PON employs the OFDM modulation scheme to increase the network capacity. As compared to TDM PON and WDM PON, OFDMA PON enjoys numerous advantages such as high speed transmission, fine granularity of bandwidth allocation, and color-free ONUs. One natural solution of the MAC protocol in

OFDM PON is to adapt current MAC protocols of TDM PONs. This is referred to as TDMA-based media access control. Taking the advantage of abundant OFDM subcarriers, OFDMA PON can use FDMA-based media access control which divides all OFDMA subcarriers into multiple nonoverlapping sets, each of which is fixedly allocated to an ONU. Another type of media access control is referred to as the hybrid TDMA and FDMA MAC protocol, which is unique for OFDMA PON. Similar to TDMA-based media access control, ONUs report their traffic information to the OLT, and the OLT allocates subcarriers to ONUs based on the real-time ONU reports. Similar to FDMA-based media access control, each ONU is dedicated with some upstream/downstream OFDMA subcarriers, which are used for control message transmission only. Performance analysis of these three MAC protocols have been compared.

Chapter 7
Hybrid Optical and Wireless Access

Optical access networking provisions high bandwidth in order to meet increasing traffic demands of end users. However, the optical solution lacks mobility, and thus limits the last mile penetration. By provisioning mobility, a rather desirable feature, to optical access, hybrid optical and wireless networks are becoming an attractive solution for wireline access network operators to expand their subscriber base [74, 99–101, 128].

7.1 Advantages of Optical Wireless Integration

Hybrid optical and wireless networks can be readily used for mobile backhaul. Global mobile networks have experienced a dramatic data rate increase in the past several years, and will continue to face the challenge of skyrocketing data rate increase in the forthcoming years owing to the growing penetration of smart phones [85]. In addition, increasingly more delay-sensitive applications such as streaming media and real-time multimedia are emerging and exerting a toll on existing mobile access networks. It was reported [32] that mobile video services would account for over 50% of the mobile services by the end of 2011. To accommodate the rapidly increasing traffic demands, mobile networks are evolving into 4G LTE or WiMAX networks whose cell tower can support up to 35 Mb/s data rate from high speed packet access (HSPA) and evolution data optimized (EV-DO). Likewise, mobile backhaul networks which connect cell towers to mobile core networks have to be accordingly upgraded to accommodate these delay-sensitive and bandwidth-demanding applications.

Generally, future mobile backhaul solutions have to meet the following four requirements. First, a low cost and high bandwidth provisioning mobile backhaul solution is needed to allow mobile service providers to operate profitably [31]. Formerly, SDH/SONET, leased lines, and microwave radio are employed to connect 2G and 3G mobile cell sites with base station controllers and radio network

N. Ansari and J. Zhang, *Media Access Control and Resource Allocation: For Next Generation Passive Optical Networks*, SpringerBriefs in Applied Sciences and Technology, DOI 10.1007/978-1-4614-3939-4_7, © The Authors 2013

controllers. To accommodate 4G traffic, using the same solution for the backhaul incurs a linear increase of the network cost with respect to the network data rate. However, mobile service providers are currently suffering from reduced average revenue per user (ARPU). To generate profitable revenue for service providers, a more cost-efficient mobile backhaul solution is strongly desired.

Second, the mobile backhaul solution should satisfy QoS requirements of a variety of applications in different wireless networks. The future radio network will accommodate a number of applications including email, voice, and multimedia video. These applications are heterogeneous in both traffic characteristics and QoS requirements. In addition, networks including 2G cellular, 3G cellular, LTE, WiFi, and WiMAX, which are deployed in the same geographical area, may share the same backhaul infrastructure for a reduced capital expenditure (CAPEX). Also, for simple network management and thus a reduced operational expenditure (OPEX), these networks may be managed over the same network control and management platform. Besides, large size macrocells, small size microcells, and picocells may coexist in future wireless networks to satisfy requirements from areas with different traffic densities; this consequently results in diversified traffic demands of different wireless access nodes. Thus, the future mobile backhaul solution should be able to satisfy respective bandwidths and QoS requirements of applications in these diversified wireless networks.

Third, the mobile backhaul solution needs to ease clock and time synchronization among base stations [12]. Base stations require accurate synchronization in order to ensure smooth handover of calls and avoid inter-cell interference. Typically, 50 parts per billion (ppb) in frequency is desired to be achieved in WCDMA/GSM systems, while no more than 3 µms delay in time is desired for CDMA2000/LTE systems. Such high requirements on synchronization discourage employment of a low-cost IP solution for mobile backhaul since IP provides best-effort services and is asynchronous in nature. Mobile backhaul solutions are desired to facilitate accurate synchronization among base stations.

Fourth, to reduce carbon footprints, out of environmental concern, mobile back-haul is desired to help reduce energy consumption of mobile cell towers and mobile core networks. Currently, energy consumption is becoming an environmental and therefore social and economic issue since one big reason of the climate change is due to the burning of fossil fuels and the direct impact of greenhouse gases on the earth's environment.

To address the low cost and synchronization challenge of mobile backhaul, carrier Ethernet studied by Metro Ethernet Forum was proposed to be employed to deliver mobile backhaul services [12]. Metro Ethernet Forum (MEF) has also identified five key attributes of carrier Ethernet from traditional LAN based Ethernet, namely, standardized services, scalability, service manageability, QoS, and reliability. MEF has also defined three categories of Ethernet virtual connections (EVC): point to point, multipoint to multipoint (ELAN), and rooted multipoint (E-Tree). These virtual connections provide Layer 2 VPN services from the access node to the IP core network. Another mobile backhaul solution is to use layer one connections such as PON technologies [50].

Formerly, TDM PON (including EPON and GPON) and WDM PON, have been proposed to connect cell towers and mobile core networks. By exploiting the large bandwidth, low power attenuation, as well as the low cost properties of PONs, this mobile backhaul solution can provide high bandwidth mobile backhaul services in a cost-efficient and energy-efficient manner.

The high bandwidth provisioning and long network reach properties of PON yield many benefits of delivering mobile backhaul services over PON. First, owing to its high capacity, PON is able to provide services to hundreds of bandwidth demanding mobile cells. Second, owing to its long reach property, PON enables the integration of metro and access networks, and thus connects the mobile core network directly with wireless access nodes. In this way, the delay of signals transmitting from the mobile core network to wireless access nodes can be readily determined. Thereby, the synchronization challenge among wireless access nodes can be easily addressed. Third, taking advantage of the low power attenuation of optical fibers, PON achieves deep penetration of optical fibers while consuming a small amount of energy.

7.2 Integrated OFDMA PON and Wireless Access

Figure 7.1 shows the OFDMA PON architecture for converged mobile and wireline access. In the proposed network, OFDMA PON delivers bandwidths to both mobile and wireline access nodes. Wireline subscribers can be business subscribers or residential subscribers, whose traffic may exhibit different behaviors. For mobile service delivery, ONUs in OFDMA PON may be integrated with 2G base stations (BSs), 3G BSs, or 4G LTE BSs. Each integrated ONU and BS is responsible for the radio resource allocation in the covered cell.

Another class of access nodes is remote radio unit (RRU), which contains the radio function in conventional BS. In future mobile networks, BS may be divided into two parts: remote radio unit (RRU) and digital baseband unit (BBU). The BBU is a centralized component performing the centralized and coordinated baseband signal processing for a set of RRUs. RRU is responsible for the radio function and antenna function, and is distributed over the cell to be nearer to end users and thus to provision high bandwidths to end users.

At the OLT side, the OLT can be connected with the 2G base station controller (BSC), 3G radio network controller (RNC), 4G LTE System Architecture Evolution Gateway (SAE GW), and 4G base station. The OLT is responsible for the allocation of resources in the OFDMA PON among various access nodes. For the case that the baseband signal processing function in 4G systems is integrated into the OLT, the OLT is responsible for the allocation of radio resources in multiple cells.

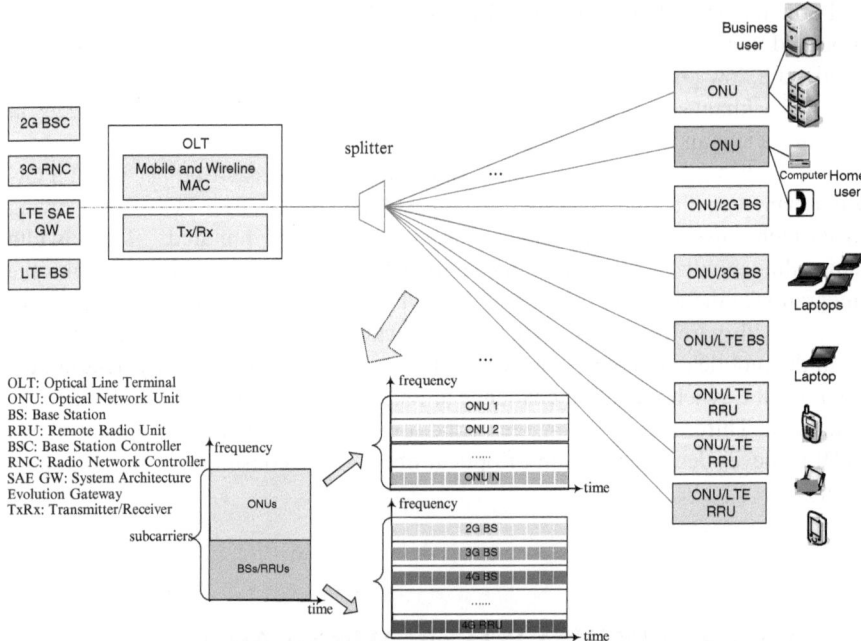

Fig. 7.1 OFDMA PON for converged mobile backhaul and wireline access

7.2.1 Resource Allocation

Resources of OFDMA PON including OFDMA subcarriers and time slots in each
OFDMA subcarrier are shared among all wireline and wireless access nodes. The
objectives of the resource allocation function are to maximize the network resource
utilization and guarantee QoS requirements of every service from each end user.

Current wireline and wireless networks provision a diversity of services ranging from low-rate delay-insensitive email services to high-rate delay-sensitive
multimedia video delivery. Moreover, traffic demands of end nodes also exhibit
diversity in terms of traffic intensity. For the wireline services, end nodes can be
business subscribers with relatively high traffic demands in the day time, or home
subscribers who generate peak traffic during the evening. For the wireless services,
the access nodes can be 2G, 3G, 4G macro cell, and 4G micro cell base stations.
Thus, the traffic demands of these wireless access nodes are rather diversified as
well. For efficient network resource utilization, the resource allocation scheme needs
to consider various QoS requirements of services and different traffic demands of
end nodes.

Typically, OFDMA PON may have thousands of OFDMA subcarriers. Taking
advantage of the abundance of OFDMA subcarriers, OFDMA PON resources can
be allocated to access nodes in two general ways. First, subcarriers are dedicated to

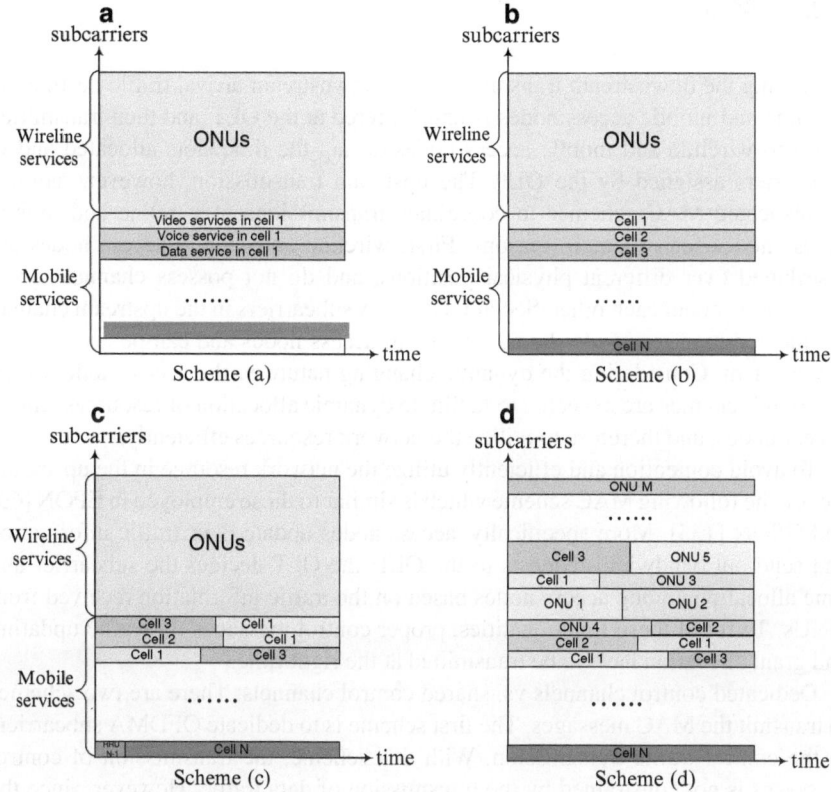

Fig. 7.2 Resource allocation of wireless access nodes

access nodes or to particular services in access nodes. Second, subcarriers are shared among services or access nodes in the TDM fashion. The first method of dedicated subcarriers does not require sophisticated scheduling to arbitrate the bandwidth allocation among access nodes or services. The second method of shared subcarriers in the TDM fashion can efficiently exploit the traffic statistical gain from different access nodes, and can thus maximize the network resource utilization.

With the focus on mobile access nodes, the following describes four types of resource allocation schemes. In Scheme (a) as shown in Fig. 7.2a, each application in a mobile access node is dedicated with a set of subcarriers. In Scheme (b) as shown in Fig. 7.2b, the access node is dedicated with a group of subcarriers, and services of the access node share these subcarriers. In Scheme (c) as shown in Fig. 7.2c, a mobile access node shares subcarriers with other mobile access nodes, but does not share OFDMA resources with optical access nodes. In Scheme (d) as shown in Fig. 7.2d, mobile access nodes share subcarriers with optical access nodes.

7.2.2 MAC

Regarding the downstream transmission, the downstream arrival traffic destined to wireline and mobile access nodes is first buffered at the OLT, and then transmitted down to wireline and mobile access nodes during the time slots allocated and in subcarriers assigned by the OLT. The upstream transmission, however, requires sophisticated MAC schemes to coordinate transmissions of wireline and mobile access nodes for two main reasons. First, wireline and mobile access nodes are distributed over different physical positions, and do not possess channel access information about each other. Second, OFDMA subcarriers in the upstream channel are the common resources shared among all access nodes and can be accessed by any of them. Considering the dynamic changing nature of the access node traffic, the MAC schemes are expected to facilitate dynamic allocation of resources among access nodes, and therefore to utilize the network resources efficiently.

To avoid contention and efficiently utilize the network resource in the upstream, we use the following MAC scheme which is similar to those employed in EPON [69] and GPON [135]. More specifically, access nodes update their traffic information and send out bandwidth requests to the OLT; the OLT decides the subcarrier and time allocation among access nodes based on the traffic information received from ONUs. To fulfill these functionalities, proper control messages for traffic updating and grant allocation have to be transmitted at the right time.

Dedicated control channels vs. shared control channels: There are two schemes to transmit the MAC messages. The first scheme is to dedicate OFDMA subcarriers to the control traffic transmission. With this scheme, the transmission of control messages is not constrained by the transmission of data traffic. However, since the control channels cannot be used for data traffic transmission even if the channel is idle, low bandwidth utilization may be resulted. The second scheme is to let control traffic share the OFDMA subcarriers with data traffic in the TDM fashion. With this scheme, a higher utilization of the network resources can be achieved as compared to the dedicated control channel scheme. However, the transmission of the control traffic such as queue status report and bandwidth allocation decision has to be carefully scheduled to avoid collision of the data traffic transmission.

Periodic control message scheduling vs. non-periodic control message scheduling: Regarding the time over which control messages are sent, control messages can be sent either periodically or nonperiodically. For the scheme of sending control messages periodically, the access node reports its queue status every fixed time duration, and the OLT sends out bandwidth allocation decisions every fixed time duration. If control traffic shares subcarriers with data traffic in the TDM manner, the bandwidth used for data traffic transmission is divided into time pieces, and the data traffic is scheduled frame by frame. Similar examples are GPON MAC schemes. For the non-periodic control message scheduling, access nodes may update their queue status upon finishing their data traffic transmission, or any time the access node or the OLT wants to send the control traffic. If the control traffic shares OFDMA subcarriers with data traffic in the TDM fashion, usually, the control packets are

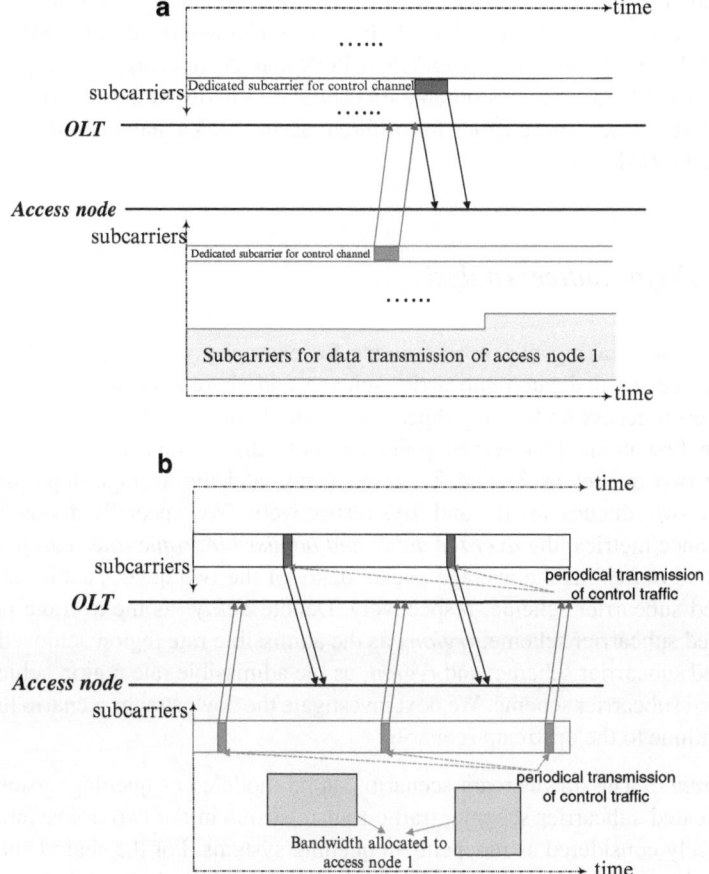

Fig. 7.3 Illustration of MAC schemes

piggybacked onto the data traffic. The EPON system uses such control messaging. If the control traffic transmission is dedicated with subcarriers, the control packets can be transmitted more flexibly and even at any time on demand.

Figure 7.3 illustrates two MAC schemes for one resource allocation. In the MAC scheme as shown in Fig. 7.3a, access nodes are dedicated with subcarriers for their respective data traffic transmission, the control traffic transmission is dedicated with subcarriers as well, and the control messages for traffic update and grant notification are sent whenever necessarily. The access node can update its traffic information to the OLT any time, and the OLT sends out its decision on the subcarriers allocated to the access node after it has made the bandwidth allocation decision. In the MAC scheme as shown in Fig. 7.3b, the access node shares OFDM subcarriers with other nodes, the control traffic shares subcarriers with data traffic in the TDM fashion, and the control traffic is sent every fixed time duration.

Different resource allocation schemes and MAC schemes yield different performances for the access nodes. Considering traffic characteristics and QoS requirements of different access nodes, OFDMA PON may design corresponding resource allocation and MAC schemes suitable for each access node. Note that different MAC and resource allocation schemes for different access nodes may be implemented in the same OFDMA PON.

7.2.3 Performance Analysis

In this section, taking two access nodes for example, we analyze and compare performances of dedicated subcarrier scheme and shared subcarrier scheme. We assume each access node is equipped with a single queue, and packets are served with the first come first served policy. Denote the average traffic arrival rates of these two queues as λ_1 and λ_2, respectively, and the average departure rates of these two queues as μ_1 and μ_2, respectively. We specially focus on two performance metrics: the *average delay and admissible traffic rate region*. Denote $delay_1^d$ and $delay_2^d$ as the average packet delay of the two queues achieved by the dedicated subcarrier scheme, respectively. Denote $delay^s$ as the average delay of the shared subcarrier scheme, $region_d$ as the admissible rate region achieved by the dedicated subcarrier scheme, and $region_s$ as the admissible rate region achieved by the shared subcarrier scheme. We next investigate the downstream scenario first, and then continue to the upstream scenario.

Downstream: The downstream scenario can be modeled as queuing systems. For the dedicated subcarrier scheme, traffic transmissions in the two access nodes are respectively considered as independent queuing systems. For the shared subcarrier scheme, the two access nodes share a single server, and can be modeled as a single queuing system. We consider three cases of different traffic arrival rates and packet size distributions.

Case 1. The packet arrival time and the packet size of both queues follow the Poisson distribution and exponential distribution, respectively.

- Dedicated subcarrier scheme: The two queues can be respectively modeled as M/M/1 queuing systems. Then, the admissible traffic region and average delays of the two queues can be obtained as follows.

$$\begin{cases} region^d = \{\lambda_1, \lambda_2, \mu_1, \mu_2 | \mu_1 > \lambda_1, \mu_2 > \lambda_2\} \\ delay_1^d = 1/(\mu_1 - \lambda_1), delay_2^d = 1/(\mu_2 - \lambda_2) \end{cases}$$

- Shared subcarrier scheme: The combined system can also be modeled as an M/M/1 queuing system with average traffic arrival rate and departure rate equaling to $\lambda_1 + \lambda_2$ and $\mu_1 + \mu_2$, respectively. Then, we can obtain the following.

$$\begin{cases} region^s = \{\lambda_1, \lambda_2, \mu_1, \mu_2 | \mu_1 + \mu_2 > \lambda_1 + \lambda_2\} \\ delay^s = 1/(\mu_1 + \mu_2 - \lambda_1 - \lambda_2) \end{cases}$$

Note that $region^s \supseteq region^d$, $delay_1^d > delay^s$, and $delay_2^d > delay^s$ if the traffic falls into the admissible rate region. Thus, the shared subcarrier scheme achieves better performances in terms of both delay and admissible rate region than the dedicated subcarrier scheme for the Poisson arrival traffic.

Case 2. The input traffic of both queues are constant bit rate (CBR) traffic, i.e., the packet arrival time of both queues are fixed, and the packet sizes of both queues are constant.

- Dedicated subcarrier scheme: Each of the two queues can be respectively modeled as an M/M/1 queuing system. Then, the admissible traffic region and average delays of the two queues can be obtained as follows.

$$\begin{cases} region^d = \{\lambda_1, \lambda_2, \mu_1, \mu_2 | \mu_1 > \lambda_1, \mu_2 > \lambda_2\} \\ delay_1^d = 1/\mu_1, delay_2^d = 1/\mu_2 \end{cases}$$

- Shared subcarrier scheme: The combined system can also be modeled as an M/M/1 queuing system with average traffic arrival rate and departure rate equaling to $\mu_1 + \mu_2$ and $\lambda_1 + \lambda_2$, respectively. Then,

$$\begin{cases} region^s = \{\lambda_1, \lambda_2, \mu_1, \mu_2 | \mu_1 + \mu_2 > \lambda_1 + \lambda_2\} \\ delay^s = 1/(\mu_1 + \mu_2) \end{cases}$$

Note that $region^s \supseteq region^d$, $delay_1^d > delay^s$, and $delay_2^d > delay^s$. Thus, the shared subcarrier scheme achieves better performances in terms of both delay and admissible rate region than the dedicated subcarrier scheme for CBR traffic.

Case 3. Queue 1 is input with CBR traffic, queue 2 is input with traffic characterized by Poisson arrival process and exponential packet size distribution, and $\lambda_1 \ll \lambda_2$, $\mu_1 \ll \mu_2$.

- Dedicated subcarrier scheme: Queue 1 can be modeled as a D/D/1 queuing system, and queue 2 can be modeled as an M/M/1 queuing system. Then, the admissible traffic region and average delays of the two queues can be obtained as follows.

$$\begin{cases} region^d = \{\lambda_1, \lambda_2, \mu_1, \mu_2 | \mu_1 > \lambda_1, \mu_2 > \lambda_2\} \\ delay_1^d = 1/\mu_1, delay_2^d = 1/(\mu_2 - \lambda_2) \end{cases}$$

- Shared subcarrier scheme: The combined queue can be modeled as a G/G/1 queuing process with average traffic arrival rate and departure rate equaling to $\lambda_1 + \lambda_2$ and $\mu_1 + \mu_2$, respectively. Owing to the assumption that $\lambda_1 \ll \lambda_2$ and $\mu_1 \ll \mu_2$, we use an M/M/1 process to approximate the G/G/1 queuing process.

Then, we can obtain:

$$\begin{cases} region^s = \{\lambda_1, \lambda_2, \mu_1, \mu_2 | \mu_1 + \mu_2 > \lambda_1 + \lambda_2\} \\ delay^s = 1/(\mu_1 + \mu_2) \end{cases}$$

Note that $region^s \supseteq region^d$ and $delay_2^d > delay^s$ if the traffic rate is within the admissible rate region. For queue 1, $delay_1^d > delay^s$ if $\mu_2 < \lambda_1 + \lambda_2$. Thus, when $\lambda_2 < \mu_2 < \lambda_1 + \lambda_2$, the dedicated subcarrier scheme yields smaller delay for queue 1 as compared to the shared subcarrier scheme; otherwise, the shared scheme achieves better performances.

Upstream: For the dedicated subcarrier scheme, the upstream queue at each access node can send the traffic over the dedicated subcarriers upon the traffic arrival. The system can be modeled as a queuing system, which is similar to the downstream scenario. Analytical results for the downstream scenario can be directly applied here.

For the shared subcarrier scheme, queues at different access nodes cannot send traffic over the shared subcarriers upon the traffic arrival since these access nodes are distributed over different physical locations and do not possess information about each other. Different from the downstream transmission case, the upstream traffic transmission requires an additional bandwidth negotiation process between the OLT and ONUs since the upstream bandwidth allocation is usually controlled by the OLT.

Here, we assume that the negotiation process between the OLT and ONUs is as follows. After packets arrive at an ONU, they are queued and wait to be reported to the OLT. The OLT receives reports from ONUs, calculates the bandwidth allocated to queued traffic of each ONU, and then sends out the grants to the ONUs. As compared to the downstream scenario, the upstream traffic experience additional delay which consists of the time between an ONU sends a report to the OLT and the same ONU receives the corresponding grant from the OLT. For notational convenience, we refer to the additional average delay as Δ.

For the above three cases of traffic arrival patterns, we obtain the following results. Note that the additional delay does not affect the admissible rate region of the network. Thus, similar to the downstream transmission scenario, the shared subcarrier scheme always admits a larger admissible traffic rate region as compared to the dedicated subcarrier scheme. Thus, here we analyze delay performances only.

Case 1. The packet arrival time and the packet size of both queues follow the Poisson distribution and exponential distribution, respectively. In this case,

$$\begin{cases} delay_1^d = 1/(\mu_1 - \lambda_1), delay_2^d = 1/(\mu_2 - \lambda_2) \\ delay^s = \Delta + 1/(\mu_1 + \mu_2 - \lambda_1 - \lambda_2) \end{cases}$$

Thus, the dedicated subcarrier scheme achieves smaller delay than the shared subcarrier scheme for queue 1 if $\Delta > 1/(\mu_1 - \lambda_1) - 1/(\mu_1 + \mu_2 - \lambda_1 - \lambda_2)$ and for queue 2 if $\Delta > 1/(\mu_2 + -\lambda_2) - 1/(\mu_1 + \mu_2 - \lambda_1 - \lambda_2)$.

Case 2. The input traffic of both queues are constant bit rate (CBR) traffic, i.e., the packet arrival time of both queues are fixed, and the packet sizes of both queues are constant. In this case,

$$\begin{cases} delay_1^d = 1/\mu_1, delay_2^d = 1/\mu_2 \\ delay^s = \Delta + 1/(\mu_1 + \mu_2) \end{cases}$$

Thus, the dedicated subcarrier scheme achieves smaller delay than the shared subcarrier scheme for queue 1 if $\Delta > 1/\mu_1 - 1/(\mu_1 + \mu_2 - \lambda_1 - \lambda_2)$ and for queue 2 if $\Delta > 1/(\mu_2 + -\lambda_2) - 1/(\mu_1 + \mu_2 - \lambda_1 - \lambda_2)$.

Case 3. Queue 1 is input with CBR traffic, queue 2 is input with traffic characterized by Poisson arrival process and exponential packet size distribution, and $\lambda_1 \ll \lambda_2$ and $\mu_1 \ll \mu_2$. In this case,

$$\begin{cases} delay_1^d = 1/\mu_1, delay_2^d = 1/(\mu_2 - \lambda_2) \\ delay^s = \Delta + 1/(\mu_1 + \mu_2 - \lambda_1 - \lambda_2) \end{cases}$$

Thus, the dedicated subcarrier scheme achieves smaller delay than the shared subcarrier scheme for queue 1 if $\Delta > 1/\mu_1 - 1/(\mu_1 + \mu_2 - \lambda_1 - \lambda_2)$ and for queue 2 if $\Delta > 1/(\mu_2 + -\lambda_2) - 1/(\mu_1 + \mu_2 - \lambda_1 - \lambda_2)$.

According to the above analysis, for the upstream transmission, the performance of the resource allocation scheme depends on Δ and the arrival traffic profile, where Δ is determined by the MAC protocol. If an access node is allocated with a set of subcarriers which can provide 1 Gb/s rate, then, it takes 1.2 μs to transmit a packet of 1,500 bytes. Even when the network is 90% loaded, the dedicated subcarrier scheme yields average delay smaller than 1.2 μs for all three cases of traffic arrival profiles.

Regarding Δ, the larger the Δ is, the more likely that the dedicated subcarrier scheme produces smaller delay than the shared subcarrier scheme. Usually, the MAC with dedicated subcarriers for control traffic produces a smaller amount of Δ than the MAC with shared subcarriers for control traffic. If a MAC compatible with GPON MAC is employed for the OFDMA PON, Δ is usually around several times that of the GPON frame length, which is defined as 1.25 μs. If a MAC similar to EPON MAC is employed for the OFDMA PON, Δ is around several times that of the EPON cycle duration, which is usually upper bounded by 2 ms. Since Δ is usually greater than the average delay produced by the dedicated subcarrier scheme, the dedicated subcarrier scheme is highly likely to achieve better delay performance than the shared subcarrier scheme in the upstream scenario.

7.3 Summary

Optical access networking provisions high bandwidth in order to meet increasing traffic demands of end users. However, the optical solution lacks mobility, and thus limits the last mile penetration. By provisioning mobility, a rather desirable feature, to optical access, hybrid optical and wireless networks are becoming an attractive solution for wireline access network operators to expand their subscriber base.

With the focus on mobile backhaul applications and OFDMA PON, this chapter discusses delivering the mobile backhaul services over OFDMA PON. Several candidate resource allocation schemes and MAC schemes have been discussed to efficiently deliver mobile services over OFDMA PON. Analytical results have been presented to compare performances of different resource allocation and MAC schemes.

Chapter 8
Green Passive Optical Networks

The future sees a clear trend of data rate increase in both wireless and wireline broadband access. These access networks may experience a dramatic increase of energy consumption in provisioning higher bandwidth as well as for other reasons [8, 28, 61]. For example, to guarantee a sufficient signal-to-noise ratio (SNR) at the receiver side for accurate recovery of high data rate signals, advanced transmitters with high transmitting signal power and advanced modulation schemes are required, thus consequently resulting in high energy consumption of the devices. Also, to provision a higher data rate, more power will be consumed by electronic circuits in network devices to facilitate fast data processing. Besides, high-speed data processing incurs fast heat buildup and high heat dissipation that further incurs high energy consumption for cooling. It is estimated that the access network energy consumption increases linearly with the provisioned data rate. It has also been reported that the LTE base station (BS) consumes more energy in data processing than the 3G UMTS systems [108, 119], and the 10 Gb/s Ethernet PON (EPON) system consumes much more energy than the 1 Gb/s EPON system. This chapter focuses on green passive optical networks,[1] the wireline aspect discoursed in great details on how to green broadband access networks [30].

The energy consumption of each access network can be split into three components: the energy consumption in the customer premises equipment (i.e., the modem), the remote node or base station (base transceiver station, BTS), and the terminal unit, which is located in the local exchange/central office. The per-customer power consumption can be expressed as: $P_a = P_{CPE} + P_{RN}/N_{RN} + 1.5P_{TU}/N_{TU}$, where P_{CPE}, P_{RN}, and P_{TU} are the powers consumed by the customer premise equipment, remote node, and terminal unit, respectively. N_{RN} and N_{TU} are the number of customers or subscribers that share a remote node and the number of customers that share a terminal unit, respectively. Table 8.1 lists the typical power consumption of these three access networks [6].

---

[1]The work presented in this chapter was supported in part by NSF under grant no. CNS-1218181.

N. Ansari and J. Zhang, *Media Access Control and Resource Allocation: For Next Generation Passive Optical Networks*, SpringerBriefs in Applied Sciences and Technology, DOI 10.1007/978-1-4614-3939-4_8, © The Authors 2013

Table 8.1 Energy consumption of ADSL, HFC, and PON

	P_{TU} (kW)	N_{TU}	P_{RN}	N_{RN}	P_{CPE}	Technology limit	Per-user capacity
ADSL	1.7	1,008	N/A	N/A	5	15 Mb/s	2 Mb/s
HFC	0.62	480	571	120	6.5	100 Mb/s	0.3 Mb/s
PON	1.34	1,024	0	32	5	2.4 Gb/s	16 Mb/s

It can be seen that PON consumes the smallest energy per transmission bit; this is attributed to the proximity of optical fibers to the end users and the passive nature of the remote node among various wireline access technologies. However, as PON is deployed worldwide, it still consumes a significant amount of energy. It is desirable to further reduce the energy consumption of PON since every single watt saving will end up with overall terawatt and even larger power saving. Reducing energy consumption of PON becomes even more important as the current PON system migrates into next-generation PON systems with increased data rate provisioning [61, 129, 135].

8.1   Reducing Energy Consumption of ONUs

The currently deployed PON systems are TDM PON systems, as shown in Fig. 8.1, in which downstream traffic is continuously broadcast to all ONUs, and each ONU transmits during the time slots in the upstream allocated by the OLT. Reducing power consumption of ONUs requires efforts across both the physical layer and MAC layer. Efforts are being made to develop optical transceivers and electronic circuits with low power consumption. Besides, multi-power mode devices with the ability of disabling certain functions can also help reduce the energy consumption of the network. However, low-power mode devices with some functions disabled may result in the degradation of the network performance. To avoid the service degradation, it is important to properly design MAC-layer control and scheduling schemes which are aware of the disabled functions. The major challenge of reducing ONUs in TDM PONs lies in the downstream transmission. In TDM PON, the downstream data traffic of all ONUs are TDM multiplexed into a single wavelength, and are then broadcasted to all ONUs. An ONU receives all downstream packets, and checks whether the packets are destined to itself. An ONU does not know when the downstream traffic arrives at the OLT, and the exact time that the OLT schedules its downstream traffic. Therefore, without a proper sleep-aware MAC control, receivers at ONUs have to be awake all the time to avoid missing their downstream packets. With the focus on the EPON, we discuss schemes in tackling the downstream challenge.

A number of schemes have been proposed to address the downstream challenge so as to reduce the energy consumption of ONUs in EPON [75, 113, 116, 131, 132, 134]. These proposed energy saving schemes can be divided into two major classes. The first class tries to design a proper MAC control scheme to convey the downstream queue status to ONUs, while the second class focuses on investigating

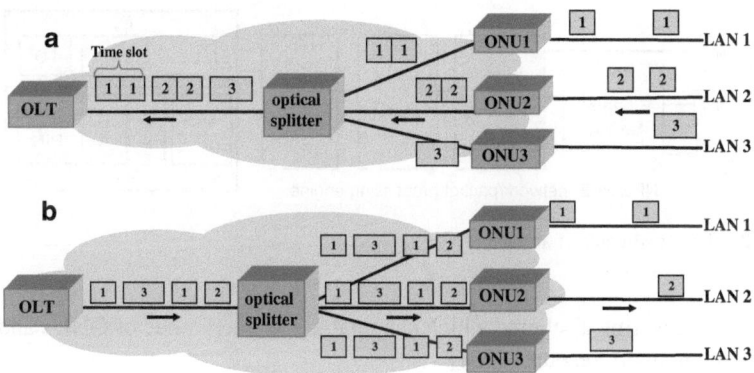

Fig. 8.1 The upstream and downstream transmission in TDM PONs

energy-efficient traffic scheduling schemes. The two-way or three-way handshake process performed between the OLT and ONUs are examples of schemes of the first class [75, 113]. Typically, the OLT sends a control message notifying an ONU that its downstream queue is empty; the ONU optionally enters the sleep mode and then sends a sleep acknowledgement or negative acknowledgement message back to the OLT. While the OLT is aware of the sleep status of ONUs, it can buffer the downstream arrival traffic until the sleeping ONU wakes up.

To implement the handshake process, the EPON MAC protocol, MultiPoint Control Protocol (MPCP) defined in IEEE 802.3ah and IEEE 802.3av, has to be modified to include new MPCP protocol data units (PDUs). In addition, the negotiation process takes at least several round trip times, implying that an ONU has to wait for several round trip times before entering the sleep status after it infers that its downstream queue is empty. This may significantly impair the energy saving efficiency.

Energy saving schemes of the second class tackle the downstream challenge by designing suitable downstream bandwidth allocation schemes. Formerly, Lee et al. [64] proposed to implement fixed bandwidth allocation (FBA) in the downstream when the network is lightly loaded. By using FBA, the time slots allocated to each ONU in each cycle are fixed and known to the ONU. Thus, ONUs can go to sleep during the time slots allocated to other ONUs. However, since traffic of an ONU dynamically changes from cycle to cycle, FBA may result in bandwidth under- or over-allocation, and consequently degrade services of ONUs to some degree. Yan et al. [117] proposed to schedule the downstream traffic and the upstream traffic simultaneously. An ONU stays in the awake status during its allocated upstream time slots, and switches into the sleep status in other time slots. Since the downstream traffic of an ONU is sent over the time slots that its upstream traffic is sent, the ONU stays in the awake status during that time period and will not miss its downstream packets. This scheme works well when traffics in the upstream and downstream are symmetric, but it may cause inefficient bandwidth utilization when the downstream traffic outweighs the upstream traffic.

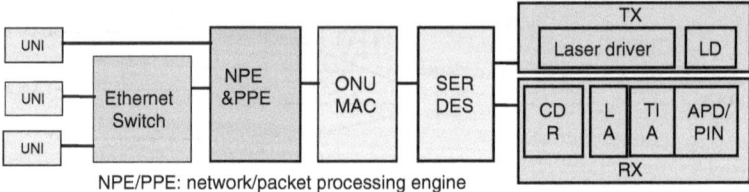

Fig. 8.2 The constituents of an ONU

We next describe a scheme which can enable the sleep mode of ONUs and best utilizes the network resource [131].

8.1.1   Sleep Status of ONUs

Figure 8.2 illustrates the constituents of an ONU. The optical module consists of an optical transmitter (Tx) and an optical receiver (Rx). The electrical module mainly contains Serializer/Deserializer (SERDES), ONU MAC, network/packet processing engine (NPE/PPE), Ethernet Switch, and user network interfaces (UNIs). When neither upstream nor downstream traffic exists, every component in the ONU can be put into "sleep". When only downstream traffic exists, the functions related to the upstream transmission can be disabled. Similarly, the functions related to receiving downstream traffic can be disabled when only upstream traffic exists. Even when the upstream traffic exists, the laser driver and laser diode (LD) do not need to be active all the time, but only during the time slots allocated to this ONU. Thus, each component in the ONU can likely "sleep", and potentially higher power saving can be achieved.

By putting each component of an ONU into sleep, an ONU ends up with multiple power levels. The "wakeup" of UNI, NPE/PPE, and switch can be triggered by the arrival of upstream traffic and the forwarding of downstream traffic from ONU MAC [51]. They are relatively easily controlled as compared to the other components. Thus, we only focus on ONU MAC, SERDES, Tx, and Rx.

8.1.2   Scenario 1: Sleep for More Than One DBA Cycle

Whether downstream/upstream traffic exists or not can be inferred based on the information of the time allocated to ONUs and queue lengths reported from ONUs, which is known to both the OLT and ONUs. If no upstream traffic arrives at an ONU, the ONU requests zero bandwidth in the MPCP REPORT message. Then, the OLT can assume that this ONU does not have upstream traffic. If no downstream traffic for an ONU arrives at the OLT, the OLT will not allocate downstream bandwidth to

---

Algorithm 6 Deciding sleeping time of ONUs

1: A:
2: **if** the Tx has not transmitted traffic for time duration of "idle_threshold" **then**
3:      $S = 1$
4:      B:
5:      Tx enters into sleep status
6:      "sleep_time"$= (2^{s-1} - 1) \times$ "short_active"$+2^{s-1} \times$ "idle_threshold"
7:      **if** "sleep_time"$> 50ms$ **then**
8:          "sleep_time"$= 50ms$
9:      **end if**
10:     Tx wakes up after sleep time duration
11:     The ONU checks the queue length and reports the queue status
12:     **if** there is queue traffic **then**
13:         Keep Tx awake
14:         $S = 0$
15:         go to A
16:     **else**
17:         $S = S + 1$
18:         go to B
19:     **end if**
20: **end if**

---

the ONU. Assume that, out of fairness concern, the OLT allocates some time slots in a dynamic bandwidth allocation (DBA) cycle to every ONU with downstream traffic. Then, considering the uncertainty of the exact time allocated to an ONU in a DBA cycle, the ONU can infer that no downstream traffic exists if it does not receive any downstream traffic within two DBA cycles.

The next question is to decide the transition between different statuses. Formerly, Kudo et al. [59] proposed periodic wakeup with the sleeping time being adaptive to the arrival traffic status. We also decide the sleeping time based on the traffic status. More specifically, we set the sleep time as the time duration that traffic stops arriving. Taking putting Tx into sleep for example, Algorithm 6 describes the sleep control scheme. We assume that Algorithm 6 is known to the OLT as well. Then, the OLT can accurately infer the time that Tx is asleep or awake.

Let "idle_threshold" be the maximum idle time duration that a transmitter stays idle before being put into sleep, "short_active" be the time taken for an ONU to check its queue status and to send out the report, and "sleep_time" be the time duration that each time an ONU sleeps. If the transmitter is idle for "idle_threshold", Tx will be put into the sleep status, and the sleep_time for the first sleep equals to "idle_threshold". Then, Tx wakes up to check its queue status and sends report to the OLT, which takes "short_active" time duration. If there is no upstream traffic being queued, Tx will enter the "sleep" status again. Until now, the elapsed time since the last time Tx transmitted data packets equals to "idle_threshold" + time duration of the first sleep + "short_active". So, for the second sleep, the sleep time duration "sleep_time" is set as "idle_threshold" + time duration of the first sleep + "short_active".

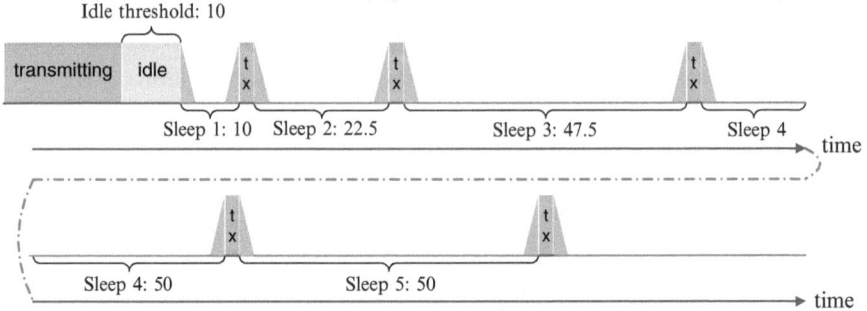

Fig. 8.3 An example of sleep time control of the transmitter

According to MPCP, ONUs send MPCP REPORT messages to the OLT every 50 ms when there is no traffic. So, we set the upper bound of "sleep_time" as 50 ms to be compatible with MPCP and also to avoid introducing too much delay of traffic arrived during the "sleep" mode. This process repeats until upstream traffic arrives. For the sth sleep, the "sleep_time" equals to "idle_threshold" + the total time duration of the former $s - 1$ sleep + $(s - 1) \cdot$ "short_active", which also equals to $(2^{s-1} - 1) \times$ "short_active"$+2^{s-1} \times$ "idle_threshold".

Figure 8.3 shows an example of the sleep time control process with "short_active" $= 2.5$ ms and "idle_threshold"$= 10$ ms. Then, the "sleep_time" of the first sleep, the second sleep, the third sleep, and the fourth sleep are as follows:

- First sleep: 10 ms
- Second sleep: "idle_threshold"$+10$ ms$+$"short_active"$= 22.5$ ms
- Third sleep: "idle_threshold"$+32.5$ ms$+2 \cdot$ "short_active"$= 47.5$ ms
- Fourth sleep: min$\{$ 50 ms, "idle_threshold"$+80$ ms$+3 \cdot$ "short_active"$\} = 50$ ms

In deciding the sleep time, "idle_threshold" and "short_active" are two key parameters which are set as follows.

- "Idle_threshold": Setting "idle_threshold" needs to consider the time taken to transit between "sleep" and "awake". Considering the transition time, the net sleep time will be reduced by the sum of the transit time from awake to sleep and the transit time from sleep to awake. Hence, "idle_threshold" should be set longer than the sum of two transit time in order to save energy in the first sleep. Currently, the time taken to power the whole ONU up is around 2–5 ms [113]. So, "idle_threshold" should be greater than 4 ms in this case. In addition, we assert that the upstream/downstream traffic queue is empty if no bandwidth is allocated to upstream/downstream traffic for "idle_threshold". To ensure this assertion is correct, "idle_threshold" should be at least one DBA cycle duration, which typically extends less than 3 ms to guarantee delay performance for some delay-sensitive service. So, Tx/Rx must sleep for over one DBA cycle with this scheme.

- "Short_active": During the short awake time of Tx, an ONU checks its upstream queue status and reports to the OLT. Hence, "short_active" should be long enough for an ONU to complete these tasks. In addition, using some upstream bandwidth for an ONU to send REPORT affects the upstream traffic transmission of other ONUs. In order to avoid the interruption of the traffic transmission of other ONUs, we set "short_active" to be at least one DBA cycle duration such that the OLT can have freedom in deciding the allocated time for an ONU to send REPORT. For Rx, during the short awake time, the OLT begins sending the queued downstream traffic if there is any. Similar to the Tx case, "short_active" is set to be at least one DBA cycle to avoid interrupting services of ONUs in the Rx case.

8.1.3   Scenario 2: Sleep Within One DBA Cycle

In the former scenario, the sleep and awake durations of Tx and Rx are greater than one DBA cycle. In this section, we discuss the scheme of putting Tx and Rx into sleep within one DBA cycle. Consider a PON with 16 ONUs. During a DBA cycle, on average, only 1/16 of time duration is allocated to an ONU. In other words, even if the upstream/downstream traffic exists, Tx/Rx only needs to be awake for 1/16 of the time, and can go to sleep for the rest 15/16 of the time. Therefore, significant energy saving can be achieved.

To enable an ONU sleep and wakeup within a DBA cycle, the transit time between awake and sleep should be less than half of the DBA cycle duration such that the next sleep time can be greater than zero, and thus energy can be saved. Wong et al. [113] reduced the transition time into as small as 1–10 ns by keeping part of the back-end circuits awake. Thus, with the advances in speeding up the transition time, it is physically possible to put an ONU into sleep within one cycle to save energy. For the upstream case, the wakeup of Tx can be triggered by ONU MAC when the allocated time comes. Tx can go to sleep after the data transmission. For the downstream case, however, it is difficult to achieve such saving since Rx does not know the time that the downstream traffic is sent and has to check every downstream packet. To address this problem, we have proposed the following sleep-aware downstream scheduling scheme [131].

For the downstream transmission, the OLT schedules the downstream traffic of ONUs one by one, and the interval between two transmissions of an ONU is determined by the sum of the downstream traffic of the other entire ONUs. Again, owing to the bursty nature of the ONU traffic, the ONU traffic in the next cycle does not vary much as compared to that in the current cycle. Accordingly, we can make an estimation of traffic of other ONUs, and put this particular ONU into sleep for some time.

More specifically, for a given ONU, denote Δ as the difference between the ending time of its last scheduled slot and the beginning time of its current scheduled slot. Then, we set the rule that the OLT will not schedule this ONU's traffic until

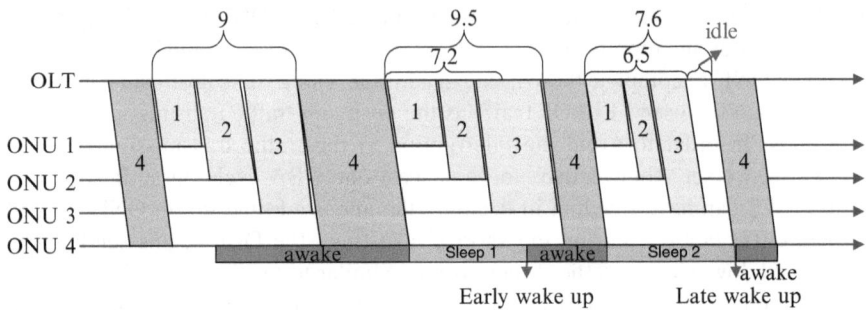

Fig. 8.4 One example of putting ONUs into sleep within one DBA cycle

$f(\Delta)$ time after the ending time of the current scheduled slot. As long as the ONU is aware of this rule, it can go to sleep for $f(\Delta)$ time duration.

Figure 8.4 illustrates one example of putting an ONU into sleep within one DBA cycle. In this example, the OLT is connected with four ONUs, and $f(\Delta)$ is set as $0.8 \cdot \Delta$. The interval between the first two scheduling of ONU 4 is 9 time units. So, the OLT will not schedule the traffic of ONU 4 until 7.2 time units later, and thus ONU 4 can sleep for 7.2 time units and then wake-up. However, this wake-up is an early wakeup since the actual transmission of the other ONUs takes 9.5 times units, which are 2.3 longer than the estimation. Similarly, the duration of the second sleep is set as 7.6 time units. However, this wake-up is a late wake-up since the actual time taken to transmit the other ONUs' traffic is 6.5 time units. The late wakeup incurs 1.1 time units of idling on the downstream channel.

As can be seen from the example, early wake-up and late wake-up are two common phenomena of this scheme. Early wake-up implies that energy can be further saved, while late wake-up results in idle time duration, and thus possibly service degradation. From a network service provider's perspective, avoiding late wake-up and the subsequent service degradation is more desirable than avoiding early wake-up. So, $0 < f(\Delta) < \Delta$ is suggested to be set. If $f(\Delta)$ is set as small as 0.5Δ, on average an ONU can still sleep $15/32$ of the time when a PON supports 16 ONUs and traffic of ONUs is uniformly distributed. Therefore, significant power saving can be achieved with this scheme.

8.2 Reducing Energy Consumption of the OLT

Formerly, the sleep mode and adaptive line rate have been proposed to efficiently reduce the power consumption of ONUs by taking advantages of the bursty nature of the traffic at the user side. It is, however, challenging to introduce the "sleep" mode into the OLT to reduce its energy consumption for the following reasons. In PONs, the OLT serves as the central access node which controls the network resource access of ONUs [73]. Putting the OLT into sleep can easily result in service disruption of ONUs in communicating with the OLT. Thus, a proper scheme is needed to reduce the energy consumption of the OLT without degrading services of end users.

8.2.1 Framework of the Energy-Efficient OLT Design

A possible solution to reduce the energy consumption of the OLT is to adapt its power-on OLT line cards according to the real-time arrival traffic. To avoid service degradation during the process of powering on/off OLT line cards, proper devices are added into the legacy OLT chassis to facilitate all ONUs in communicating with power-on line cards.

In the central office, the OLT chassis typically comprises multiple OLT line cards, each of which communicates with a number of ONUs. In the currently deployed EPON and GPON systems, one OLT line card usually communicates with either 16 or 32 ONUs. To avoid service disruptions of ONUs connected to the central office, all these OLT line cards in the OLT chassis are usually powered on all the time. To reduce the energy consumption of the OLT, our main idea is to adapt the number of power-on OLT line cards in the OLT chassis to the real-time incoming traffic.

Denote C as the provisioned data rate of one OLT line card, L as the total number of OLT line cards, N as the number of ONUs connected to the OLT chassis, and $r^i(t)$ as the arrival traffic rate of ONU i at time t. By powering on all the OLT line cards, the overall data rate accommodated by the OLT chassis equals to $C * L$, which may be greater than the real-time incoming traffic, i.e., $\sum_{i=1}^{N} r^i(t)$. Denote l as the smallest number of required OLT line cards which can provision at least $\sum_{i=1}^{N} r^i(t)$ data rate. Then, our ultimate objective is to power on only l OLT line cards to serve all N ONUs at a given time t instead of powering on all L line cards. However, powering off OLT line cards may result in service disruptions of ONUs communicating with these OLT line cards. To avoid service disruption, power-on OLT line cards should be able to provision bandwidth to all ONUs connected to the OLT chassis. To address this issue, we have proposed several modifications over the legacy OLT chassis to realize the dynamic configuration of the OLT as will be presented next [136].

8.2.2 The OLT with Optical Switch

To dynamically configure the communications between OLT line cards and ONUs, one scheme we propose is to place an optical switch in front of all OLT line cards as shown in Fig. 8.5a. The function of the optical switch is to dynamically configure the connections between OLT line cards and ONUs. When the network is heavily loaded, the switches can be configured such that each PON system communicates

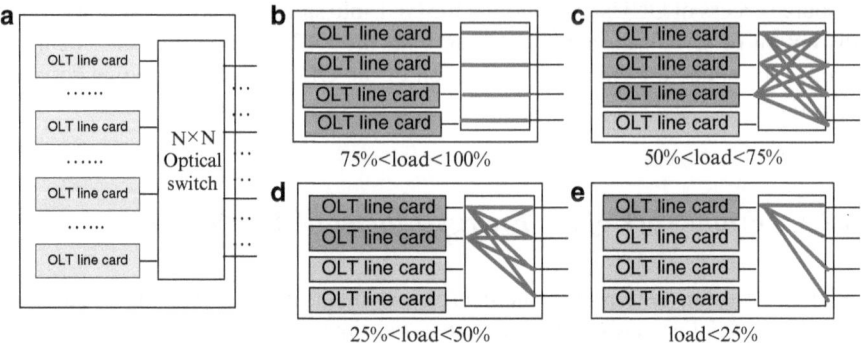

Fig. 8.5 The OLT with optical switches

with one OLT line card. When the network is lightly loaded, the switches can be configured such that multiple PON systems communicate with one line card. Then, some OLT line cards can be powered off, thus saving energy consumption.

Assume the energy consumption of the optical switch is negligible. As compared to the scheme of always powering on all L line cards, the scheme of powering on only $\lceil \sum_{i=1}^{N} r^i(t)/C \rceil$ line cards can achieve relative energy saving as large as

$$1 - \frac{\lceil \sum_{i=1}^{N} r^i(t)/C \rceil}{L}.$$

Then, the average energy saving over time span T equals to

$$\frac{\int_{t=0}^{T} 1 - \lceil \sum_{i=1}^{N} r_i(t)/C \rceil / L}{T}.$$

Figure 8.5b–e illustrates the configuration of switches for the case that one OLT chassis contains four OLT line cards. Define traffic load as $\lceil \sum_{i=1}^{N} r^i(t)/(L \cdot C) \rceil$, where $\lceil \sum_{i=1}^{N} r^i(t)/C \rceil$ is the total arrival traffic rates of all ONUs and $L \cdot C$ is the capacity provisioned by all OLT line cards. By dynamically configuring switches, the number of power-on OLT line cards is reduced from four to x when the traffic load falls between $x/4$ and $(x+1)/4$. Thus, a significant amount of power can be potentially saved.

However, switches take time to change configurations. The switch reconfiguration time may affect the ONU performances when powering on/off OLT line cards. We investigate the impacts of the switch reconfiguration time on EPON and GPON, respectively.

- EPON: We argue that services of EPON ONUs are not affected when the switch configuration time is as large as 50 ms.

In EPON, the upstream bandwidth allocation is controlled by the OLT. ONUs transmit the upstream traffic using the allocated time durations indicated in the GATE message. IEEE 802.3ah [44] specifies that ONUs need to send GATE messages every 50 ms to maintain registration even if they do not have traffic to transmit. Thus, the disrupted servicing time can be transparent to EPON ONUs when the switch reconfiguration time is as large as 50 ms, which can be satisfied by most optical switches.

- GPON: We argue that services of GPON ONUs are not affected when the switching time is not greater than 125 μs.

For GPON, ITU-T G.984.3 [46] specifies a fixed frame length of 125 μs. An ONU receives downstream control messages and sends its upstream data or control traffic every GPON frame, i.e., 125 μs. Therefore, services of GPON ONUs are not affected when the switch configuration time is not greater than 125 μs.

8.2.3   The OLT with Cascaded 2×2 Switches

Another problem with optical switches is their high costs. The prices of optical switches vary from their manufacturing techniques. At present, there are generally four kinds of optical switches: opto-mechanical switches, micro-electro-mechanical system (MEMS), electro-optic switches, and semiconductor optical amplifier switches. Currently, the opto-mechanical switches are less expensive than the other three kinds. Simply because of their low prices, opto-mechanical switches are generally the adopted choices in designing energy-efficient OLT.

For opto-mechanical switches, an important constraint is their limited port counts. Popular sizes of opto-mechanical switches are 1×2 and 2×2. Considering the port count constraints, the cascaded 2×2 switches structure can be used to achieve the dynamic configuration of the OLT. More specifically, the cascaded 2×2 switch containing $\log_2 N$ stages and $(N-1)$ 2×2 switches can replace an $N \times N$ switch. Figure 8.6a illustrates the cascaded switches. In the switch, the kth stage contains 2^{k-1} switches.

Figure 8.6b shows a two-stage cascaded 2×2 switch to replace a 4×4 switch. As illustrated in Fig. 8.6c–e, when the traffic load is greater than 50%, one PON system is connected with one OLT line card; when the traffic load is between 25 and 50%, two PON systems are connected with one OLT line card; when the traffic load is less than 25%, all PON systems are connected with a single OLT line card.

Here, we analyze the saved energy of the OLT equipped with cascaded switches. Assume the traffic is uniform among all ONUs. Then, when the traffic load is between 50 and 100%, all OLT line cards are powered on; when the traffic load is between 25 and 50%, half of OLT line cards are powered on. Generally, when the traffic load is between $1/2^k$ and $1/2^{k+1}$, $1/2^k$ of OLT line cards are powered on.

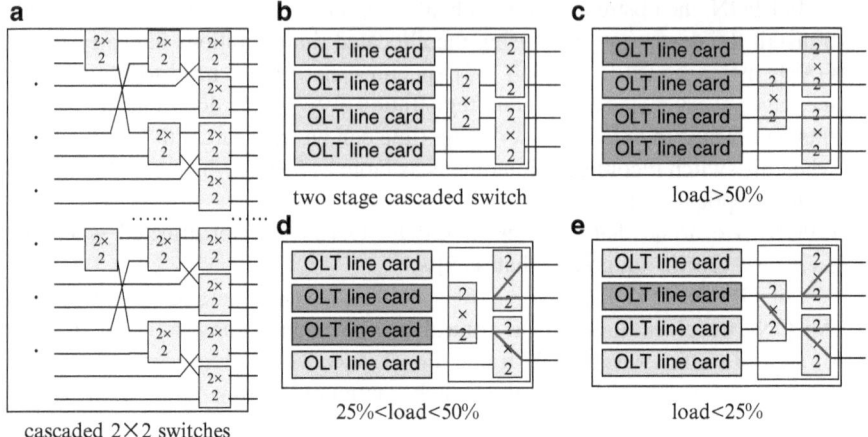

Fig. 8.6 The OLT with multistage cascaded switches

Then, the saved energy equals to $1 - 1/2^{\lfloor log_2 1/load \rfloor}$, where $load = \lceil \sum_{i=1}^{N} r^i(t)/(L \cdot C) \rceil$ as defined before. As compared to the OLT with an $N \times N$ switch, the OLT with cascaded 2×2 switches saves a less amount of energy.

8.3   Summary

This chapter analyzes and compares energy consumption of existing broadband wireline access networks. We have also summarized and discussed various techniques which have been proposed to reduce energy consumption of PON access networks. We have compared energy consumption of various broadband access solutions, and discussed several greening techniques such as enabling the sleeping mode to reduce energy consumption of ONUs and adapting line rates to reduce energy consumption of the OLT.

Chapter 9
Looking Forward

By now, the readers should have walked through the journey with us via the past chapters observing how PON has evolved from various flavors of TDM PON (from the initial concoction of APON to currently deployed BPON, EPON, and GPON) to WDM PON with potential huge bandwidth provisioning; to OFDM PON with the benefits of high speed transmission, finer granularity of bandwidth provisioning, and color-free ONUs; to hybrid optical and wireless integration in provisioning mobility; and to green PON in the effort to reduce information and communication technology (ICT) carbon footprints. Currently deployed single-channel TDM PON systems will not likely be able to meet the ever growing traffic demands, both in quantity and variety, in the future. We will quickly recap what have transpired from the previous chapters in looking forward to what will likely actualize in the future. The PON evolution has continued, initially, in two stages: mid-term and long-term. The basic requirement of the evolution is achieving higher bandwidth, defined as NG-PON1 and NG-PON2, respectively, by the GPON interest group of the Full Service Access Network (FSAN). Specifically, FSAN has decided that the mid-term NG-PON1 should coexist with the currently deployed GPON systems and reuse the outside plant based on the current optical components and cost control. That is, the main requirement of NG-PON1 is compatibility. Besides, NG-PON1 was specified to provide even larger power budget so as to achieve increased split ratio and reach distance. In order to achieve these goals, NG-PON1 adapts advanced optical devices. The standardization of the asymmetric bandwidth edition of NG-PON1 was defined in XG-PON1, which was published in 2010 by FSAN and ITU-T in the ITU-T G.987 series [47]. It provides 10G downstream/2.5G upstream data rate. XG-PON2, another enhanced version for GPON mid-term evolution, was recommended to provide 10G/10G symmetric bandwidth and is not yet standardized (at the time of this writing). For the long-term development, NG-PON2 is still under discussion. Unlike NG-PON1, NG-PON2 can be an entirely new PON technology without considering the legacy PON compatibility constraints but with the overmatching of the legacy PON in terms of bandwidth, distance, security, and some other aspects.

N. Ansari and J. Zhang, *Media Access Control and Resource Allocation: For Next Generation Passive Optical Networks*, SpringerBriefs in Applied Sciences and Technology, DOI 10.1007/978-1-4614-3939-4_9, © The Authors 2013

In evolving from NG-PON1 to NG-PON2, more technologies can be adopted into this long-term evolution.

Meanwhile, IEEE EFM has made substantial effort in standardizing 10G EPON [44], the successor of 1G EPON. The specifications of 10G EPON, IEEE 802.3av, were completed at the end of 2009. Upgrades are mainly focused on the physical layer. Unlike IEEE 802.3ah, the 1G EPON specifications, that adapts 8b/10b line code, IEEE 802.3av chooses 64b/66b line code, and supports 10G downstream/1G upstream asymmetric data rate and 10G/10G symmetric data rate. In addition, it defines mandatory specific FEC function to increase power budget. 10G EPON is also a TDM PON without any change to TDM EPON MPCP, and it is also compatible with the current 1G EPON system.

Both XG-PON1 and 10G EPON are 10G PONs which provide 10 Gb/s downstream line rate and no more than 10 Gb/s upstream line rate. They have been going through the field trial stage recently. However, the cost of optical devices for 10G PON is almost more than double that of traditional PON at the time of this writing, while the 10G bandwidth is far from meeting today's demand, especially in the upstream, in which bandwidth is still shared among multiple end-users by timeslots. A comparison of legacy PON, 10G PON, and NG-PON is illustrated in Table 9.1.

In choosing the technologies for NG-PONs, industries should consider the need of customers as well as that of network providers. Basically, customers want to achieve good quality of experience at the lowest possible cost while network providers seek to maximize revenues by supporting attractive services. The basic requirement of the next generation PON is providing end users with higher bandwidth at even lower cost. The main system requirements can be summarized as follows:

- Bandwidth: Bandwidth-sensitive services are getting popular. Providers are seeking to support higher per-user data rate which is likely to reach 20 Mb/s soon and 100 Mb/s in the near future.
- Cost: Cost is always the principal consideration for both the service providers and the end users. Typically, there are two methods to achieve cost efficiency: (1) protection of the current investment such as imposing compatibility with traditional PONs like the requirement of XG-PON1 and 10G EPON, and (2) high resource utilization such as statistical multiplexing of bandwidth, wavelength, and/or subcarriers.
- Reach extension: Long-reach and large-split PON is attractive since it extends the coverage span of PON that can potentially integrate metro and access networks into a single platform, and simplify the management of networks by reducing center nodes such as OLTs and saving energy.
- Scalability: NG-PON should have the capability to accommodate the ever growing number of customers without sacrificing network performance.
- Reliability: NG-PON should cover more customers than what are covered today, and it is thus critical to prevent fiber failure or nodes collapse. The reliability of NG-PON should be of paramount importance.

Table 9.1 Comparison of PONs

	APON/BPON	GPON	XG-PON1	EPON	10G-EPON
Standard	ITU-T G.983	ITU-T G.984	ITU-T G.987	IEEE 802.3ah	IEEE 802.3av
Downstream speeds	622 Mbps	2.488 Gbps	9.9528 Gbps	1.25 Gbps	10.3125 Gbps
Upstream speeds	155 Mbps	1.244 Gbps	2.488 Gbps, 9.9528 Gbps	1.25 Gbps	1.25 Gbps, 10.3125 Gbps

- Energy efficiency: Access networks consume the majority of the ICT energy [6] owing to the large quantity of access nodes, each of which requires a non-negligible amount of power supply even in the "standby" mode. As broadband access is deployed world-wide, every single watt increase will end up with even terawatts overall power increase. Therefore, it is vital to reduce the energy consumption of PONs, being a dominant choice of access networks, as much as possible.

In order to meet the above requirements for next-generation PON, first in the physical layer or architecture perspective, several candidate multiple access technologies for NG-PON have emerged. The main candidates are WDM PON, OFDMA PON, and OCDMA PON. Second, issues consequentially arisen from these fundamental architectures in order to satisfy customers and providers of NG PONs must be addressed. Regardless of the adopted access techniques, pressing issues to be addressed include dynamic resource allocation (DRA), energy efficiency, network integration, and service expansion.

As an alternative to WDM PON (Chap. 5) and OFDM PON (Chap. 6), Optical Code Division Multiple Access (OCDMA) has been considered as a viable access technique for next-generation PON [24, 53, 118]. In the OCDMA-PON network, an optical coder/encoder is equipped at the transmitter and receiver, respectively. Multiple users share the same transmission media by encoding different users' data into different optical codes. OCDMA PON is uniquely featured with its full asynchronous transmission which can potentially simplify network management as well as promote flexibility of QoS control. In order to achieve cost-efficiency, Yang [118] proposed to increase the number of codes, while Gharaei et al. [24] investigated hybrid OCDMA-TDM PON with colorless ONUs and Kataoka et al. [53] looked into hybrid OCDMA-WDM PON with multi-port optical encoder/decoders to support terabit capacity and long reach distance. However, the multiple access interference (MAI) caused by the crosstalk between different users sharing the same channel and the beat noise inherited to coherent OCDMA remain big challenges, which limit the number of simultaneous users of the system.

These technologies for NG-PON are still under development in research labs, and years away from large scale deployment owing to technical complexities, high component device cost, and dismal industry chain maturity. Nevertheless, the following pressing issues will continue to draw much attention and effort.

- Dynamic resource allocation remains critical for NG-PONs to achieve better resource utilization. Resource contention is no longer limited to timeslots exploited for TDM PON but also includes wavelength channels for WDM/TDM PON, subcarriers for OFDM PON, and varied resources for different scenarios. Voluminous research results in advancing hybrid WDM/TDM PON [79, 94, 110, 130, 133], OFDMA PON [25, 51, 111, 115, 137], and LR-PON [49, 103] are continuously to be made.
- Chapter 8 has presented various proposals to reducing energy consumption of ONUs and OLTs. On a bigger picture, the future sees a clear trend of data rate increase in both wireline and wireless access. These access networks may

experience a dramatic increase of energy consumption in provisioning higher bandwidth as well for other purposes. Thus, enhancing energy efficiency will continue to preoccupy much research effort in greening PONs, as part of the overall effort in Greening At The Edges (GATE) [22].

- Optical networks are traditionally considered proficient in provisioning high bandwidth and long distance coverage. By facilitating mobility, hybrid optical and wireless access, as discoursed in Chap. 7, is becoming a viable solution for network operators to augment their subscriber base and generate revenues. Moreover, a powerful and versatile platform, as a future mobile backhaul with extended coverage span as well as provisioned mobility, can be built by employing long-reach and large-split PON in this hybrid optical and wireless access. To achieve such seamless convergence, many challenges have to be overcome, including new architectures [67], MAC protocols, dynamic wavelength, bandwidth and subcarrier allocation (DWBSA), and routing and load balancing [15].

- Cloud computing, "the use of computing resources (hardware and software) that are delivered as a service over a network" (http://en.wikipedia.org/wiki/cloud_computing), is becoming the game changer on how services are delivered. Cloud services are basically provisioned to end-users through several centrally managed big data centers located at the edges of the core networks by major players such as Amazon, Google, Microsoft, and IBM. However, these service providers do not own the networks. As a result, they cannot guarantee user quality of experience (QoE) by intelligent management especially at the networking level. Moreover, the latency of WAN normally is over 150 ms which is intolerable for most of the services in this centralized model. This potentially bring business opportunities to telco-carriers who own the broadband access networks and can take the cloud pools much closer to the customers so as to guarantee QoS [106] at both the cloud and networking levels. Naturally, can telco-carriers, which have made tremendous investment in deploying their PON systems, capitalize on their CAPEX by providing value-added services such as cloud services? Though not much have been reported in the open literature [72, 95, 96], many new research and development findings on PON cloud will be reported in the near future.

PON has been demonstrated to be a future-proof solution for broadband access. As long as there are demands for capacities and services in accommodating various bandwidth-hungry applications, breakthroughs in PON technologies in terms of optical devices, issues within the Open Systems Interconnection (OSI) layered architecture, access techniques, and value-added services, are forthcoming.

References

1. http://en.wikipedia.org/wiki/Internet_traffic (viewed on 12/26/2012)
2. http://www.cisco.com/en/US/netsol/ns827/networking_solutions_sub_solution.html#~ forecast (viewed on 12/26/2012)
3. An, F.T., Gutierrez, D., Kim, K.S., Lee, J.W., Kazovsky, L.: SUCCESS-HPON: A next-generation optical access architecture for smooth migration from TDM-PON to WDM-PON. IEEE Comm. Mag. **43**(11), S40–S47 (2005). doi: 10.1109/MCOM.2005.1541698
4. Assi, C., Ye, Y., Dixit, S., Ali, M.: Dynamic bandwidth allocation for quality-of-service over Ethernet PONs. IEEE J. Sel. Areas Comm. **21**(9), 1467–1477 (2003). doi: 10.1109/ JSAC.2003.818837
5. Bai, X., Shami, A., Ghani, N., Assi, C.: A hybrid granting algorithm for QoS support in Ethernet passive optical networks. In: 2005 IEEE International Conference on Communications, IEEE Conference Publications, Piscataway, NJ, **3**, 1869–1873 (2005)
6. Baliga, J., Ayre, R., Hinton, K., Tucker, R.: Energy consumption in wired and wireless access networks. IEEE Comm. Mag. **49**(6), 70–77 (2011)
7. Banerjee, A., Park, Y., Clarke, F., Song, H., Yang, S., Kramer, G., Kim, K., Mukherjee, B.: Wavelength-division-multiplexed passive optical network (WDM-PON) technologies for broadband access: A review. J. Opt. Netw. **4**(11), 737–758 (2005)
8. Bianco, C., Cucchietti, F., Griffa, G.: INTELEC 07 - 29th International Telecommunications Energy Conference. In: Proceedings of a meeting held Rome, Italy, **2**, pages 927, Institute of Electrical and Electronics Engineers (IEEE), Piscataway, NJ (2007)
9. Tanaka, K., Agata, A. and Horiuchi Y.: IEEE 802.3av 10G-EPON Standardization and Its Research and Development Status, IEEE/OSA J. Lightwave Tech., **28**(4), 651–61 (2010)
10. Bonilla, M., Barbosa, F., Moschim, E.: Techno-economical comparison between GPON and EPON networks. In: ITU-T Kaleidoscope: Innovations for Digital Inclusions. IEEE Conference Publications, Piscataway, NJ (2009)
11. Boyd, S., Vandenberghe, L.: Convex Optimization. Cambridge University Press, Cambridge (2004)
12. Briggs, P., Chundury, R., Olsson, J.: Carrier ethernet for mobile backhaul. IEEE Comm. Mag. **48**(10), 94–100 (2010)
13. Centeno, G., Armacost, R.L.: Parallel machine scheduling with release time and machine eligibility restrictions. Comput. Ind. Eng. **33**(3–4), 273–276 (1997). doi: 10.1109/ MCOM.2007. 344582
14. Centeno, G., Armacost, R.L.: Minimizing makespan on parallel machines with release time and machine eligibility restrictions. Int. J. Prod. Res. **42**(6), 1243–1256 (2004)

N. Ansari and J. Zhang, *Media Access Control and Resource Allocation: For Next Generation Passive Optical Networks*, SpringerBriefs in Applied Sciences and Technology, DOI 10.1007/978-1-4614-3939-4, © The Authors 2013

15. Chowdhury, P., Tornatore, M., Sarkar, S., Mukherjee, B.: Building a green wireless-optical broadband access network (WOBAN). IEEE/OSA J. Lightwave Tech. **28**(16), 2219–2229 (2010)
16. Coffman, E., Garey, M., Johnson, D.: An application of bin-packing to multiprocessor scheduling. SIAM J. Comput. **7**(1), 1–17 (1978)
17. Dhaini, A.R., Assi, C.M., Maier, M., Shami, A.: Dynamic wavelength and bandwidth allocation in hybrid TDM/WDM EPON networks. IEEE/OSA J. Lightwave Tech. **25**(1), 277–286 (2007)
18. Effenberger, F., Clearly, D., Haran, O., Kramer, G., Li, R.D., Oron, M., Pfeiffer, T.: An introduction to PON technologies. IEEE Comm. Mag. **45**(3), S17–S25 (2007). doi: 10.1109/MCOM.2007.344582
19. Fiedler, M., Hossfeld, T., Tran-Gia, P.: A generic quantitative relationship between quality of experience and quality of service. IEEE Netw. **24**(2), 36–41 (2010). doi: 10.1109/MNET.2010.5430142
20. Full-service Access Network (FSAN) Group. http://www.fsan.org/
21. Garey, M., Johnson, D.: Computers and Intractability: A Guide to the Theory of NP-Completeness. W.H. Freeman, New York (1979)
22. GATE: Greening at the edges. NSF Project under grant no. CNS-1218181, August 1, 2012 to July 31, 2015 (PI: Nirwan Ansari)
23. Ghani, N., Shami, A., Assi, C., Raja, M.: Intra-ONU bandwidth scheduling in Ethernet passive optical networks. IEEE Comm. Lett. **8**(11), 683–685 (2004). doi: 10.1109/LCOMM.2004.837664
24. Gharaei, M., Lepers, C., Gallion, P.: Upstream OCDMA-TDM passive optical network using a novel sourceless ONU. In: 17th European Conference on Networks and Optical Communications (NOC). IEEE Conference Publication, Piscataway, NJ (2012)
25. Giacoumidis, E., Wei, J., Yang, X., Tsokanos, A., Tang, J.: Adaptive-modulation-enabled WDM impairment reduction in multichannel optical OFDM transmission systems for next-generation PONs. IEEE Photonics J. **2**(2), 130–140 (2010)
26. Graham, R., Lawler, E., Lenstra, J., Rinnooy Kan, A.: Optimization and approximation in deterministic sequencing and scheduling: A survey. Ann. Discrete Math. **5**, 287–326 (1979)
27. Grobe, K., Elbers, J.P.: PON in adolescence: From TDMA to WDM-PON. IEEE Comm. Mag. **46**(1), 26–34 (2008). doi: 10.1109/MCOM.2008.4427227
28. Gupta, M., Singh, S.: Greening of the internet. In: Proceedings of the ACM SIGCOMM Conference on Applications, Technologies, Architectures, and Protocols for Computer Communication, Karlsruhe, Germany. ISBN 1-58113-735-4, ACM, New York, 25–29 (2003)
29. Hajduczenia, M., da Silva, H.J., Monteiro, P.P.: EPON versus APON and GPON: A detailed performance comparison. OSA J. Opt. Netw. **5**(4), 298–319 (2006)
30. Han, T., Zhang, J., Ansari, N.: Chapter 17: Green broadband access networks. In: Obaidat, M.S., Anpalagan, A., Woungang, I. (eds.) Handbook of Green Information and Communication Systems. Academic, New York (2013)
31. Harno, J.: Impact of 3G and beyond technology development and pricing on mobile data service provisioning, usage and diffusion. Telemat. Inform. **27**(3), 269–282 (2010)
32. Heikkinen, M.V.J., Berger, A.W.: Comparison of user traffic characteristics on mobile-access versus fixed-access networks. In: Proceedings of the 13th International Conference on Passive and Active Measurement, PAM'12, pp. 32–41. Springer, Berlin (2012)
33. Heron, R.W., Pfeiffer, T., van Veen, D.T., Smith, J., Patel, S.S.: Technology innovations and architecture solutions for the next-generation optical access network. Technical Report 1, Bell Labs Technical Journal (2008)
34. Hiertz, G., Denteneer, D., Stibor, L., Zang, Y., Costa, X., Walke, B.: The IEEE 802.11 universe. IEEE Comm. Mag. **48**(1), 62–70 (2010)
35. Honcharenko, W., Kruys, J., Lee, D., Shah, N.: Broadband wireless access. IEEE Comm. Mag. **35**(1), 20–26 (1997)
36. Hoßfeld, T., Binzenhöfer, A.: Analysis of Skype VoIP traffic in UMTS: End-to-end QoS and QoE measurements. Comput. Netw. **52**(3), 650–666 (2008)

37. Hou, E., Ansari, N., Ren, H.: A genetic algorithm for multiprocessor scheduling. IEEE Trans. Parallel Distrib. Syst. **5**(2), 113–120 (1994)
38. http://www.itu.int/rec/T-REC-I.113-199706-I/e (1997)
39. http://www.itu.int/rec/T-REC-J.112/en (1998)
40. http://www.itu.int/rec/T-REC-J.122/en (2004)
41. http://www.itu.int/rec/T-REC-G.992.5/en (2005)
42. http://www.telecompaper.com/news/global-ftthb-subscribers-reach-67-mln-at-mid-2011 (2012)
43. Hutcheson, L.: FTTx: Current status and the future. IEEE Comm. Mag. **46**(7), 90–95 (2008). doi: 10.1109/MCOM.2008.4557048
44. IEEE approved Draft Std P802.3av/D3.4 (2009)
45. ITU-T: Recommendation G. 1010: End-user multimedia QoS categories (2001)
46. ITU-T Recommendation G.984 series: http://www.itu.int/rec/T-REC-G/e
47. ITU-T Recommendation G.987 series: http://www.itu.int/rec/T-REC-G.987/en
48. Jiang, S., Xie, J.: A frame division method for prioritized DBA in EPON. IEEE J. Sel. Areas Comm. **24**(4), 83–94 (2006). doi: 10.1109/JSAC.2006.1613774
49. Jiménez, T., Merayo, N., Fernández, P., Durán, R., de Miguel, I., Lorenzo, R., Abril, E.: Implementation of a PID controller for the bandwidth assignment in long-reach PONs. IEEE/OSA J. Opt. Comm. Netw. **4**(5), 392–401 (2012)
50. Kani, J., Bourgart, F., Cui, A., Rafel, A., Campbell, M., Davey, R., Rodrigues, S.: Next-generation PON-Part I: Technology roadmap and general requirements. IEEE Comm. Mag. **47**(11), 43–49 (2009)
51. Kanonakis, K., Giacoumidis, E., Tomkos, I.: Physical-layer-aware MAC schemes for dynamic subcarrier assignment in OFDMA-PON networks. IEEE/OSA J. Lightwave Tech. **30**(12), 1915–1923 (2012)
52. Kanonakis, K., Tomkos, I., Pfeiffer, T., Prat, J., Kourtessis, P.: ACCORDANCE: A novel OFDMA-PON paradigm for ultra-high capacity converged wireline-wireless access networks. In: Transparent Optical Networks (ICTON), 12th International Conference on, pp. 1–4, IEEE Conference Puboication, Piscataway, NJ (2010). doi: 10.1109/ICTON.2010.5549027
53. Kataoka, N., Wada, N., Cincotti, G., Kitayama, K.: 2.56 tbps (40-gbps× 8-wavelength× 4-oc× 2-pol) asynchronous WDM-OCDMA-PON using a multi-port encoder/decoder. In: 37th European Conference and Exhibition on Optical Communication (ECOC). IEEE Conference Publications, Piscataway, NJ (2011)
54. Kim, C., Yoo, T.W., Kim, B.T.: A hierarchical weighted round robin EPON DBA scheme and its comparison with cyclic water-filling algorithm. IEEE Int. Conf. Comm. 2156–2161 (2007). doi: 10.1109/ICC.2007.363
55. Kim, K.: On the evolution of PON-based FTTH solutions. Inf. Sci. **149**(1), 21–30 (2003)
56. Kramer, G., Mukherjee, B., Dixit, S., Ye, Y., Hirth, R.: Supporting differentiated classes of service in Ethernet passive optical networks. J. Opt. Netw. **1**(8), 280–298 (2002)
57. Kramer, G., Mukherjee, B., Pesavento, G.: IPACT a dynamic protocol for an Ethernet PON (EPON). IEEE Comm. Mag. **40**(2), 74–80 (2002). doi: 10.1109/35.983911
58. Kramer, G., Pesavento, G.: Ethernet passive optical network (EPON): Building a next-generation optical access network. IEEE Comm. Mag. **40**(2), 66–73 (2002). doi: 10.1109/35.983910
59. Kubo, R., Kani, J., Ujikawa, H., Sakamoto, T., Fujimoto, Y., Yoshimoto, N., Hadama, H.: Study and demonstration of sleep and adaptive link rate control mechanisms for energy efficient 10G-EPON. IEEE/OSA J. Opt. Comm. Netw. **2**(9), 716–729 (2010)
60. Kwong, K.H., Harle, D., Andonovic, I.: Dynamic bandwidth allocation algorithm for differentiated services over WDM EPONs. In: The Ninth International Conference on Communications Systems, Optical Society of America 2010 Massachusetts Ave., N.W. Washington, D.C. 20036-1012. pp. 116–120, USA (2004). doi: 10.1109/ICCS.2004.1359350

61. Lange, C., Braune, M., Gieschen, N.: On the energy consumption of FTTB and FTTH access networks. In: National Fiber Optic Engineers Conference, Optical Society of America 2010 Massachusetts Ave., N.W. Washington, D.C. 20036–1012 USA (2008)

62. Lee, C.H., Lee, S.M., Choi, K.M., Moon, J.H., Mun, S.G., Jeong, K.T., Kim, J.H., Kim, B.: WDM-PON experiences in Korea. J. Opt. Netw. **6**(5), 451–464 (2007)

63. Lee, C.H., Sorin, W.V., Kim, B.Y.: Fiber to the home using a PON infrastructure. IEEE/OSA J. Lightwave Tech. **24**(12), 4568–4583 (2006). doi: 10.1109/JLT.2006.885779

64. Lee, S., Chen, A.: Design and analysis of a novel energy efficient ethernet passive optical network. In: Ninth International Conference on Networks (ICN), pp. 6–9. IEEE Conference Publications, Piscataway, NJ (2010)

65. Lee, S.M., Mun, S.G., Kim, M.H., Lee, C.H.: Demonstration of a long-reach DWDM-PON for consolidation of metro and access networks. IEEE/OSA J. Lightwave Tech. **25**(1), 271–276 (2007). doi: 10.1109/JLT.2006.887179

66. Lenstra, J., Shmoys, D., Tardos, E.: Approximation algorithms for scheduling unrelated parallel machines. Math. Program. **46**, 259–271 (1990)

67. Li, Y., Wang, J., Qiao, C., Gumaste, A., Xu, Y., Xu, Y.: Integrated fiber-wireless (FiWi) access networks supporting inter-ONU communications. IEEE/OSA J. Lightwave Tech. **28**(5), 714–724 (2010)

68. Limb, J., Sala, D.: A protocol for efficient transfer of data over hybrid fiber/coax systems. IEEE/ACM Trans. Netw. **5**(6), 872–881 (1997)

69. Luo, Y., Ansari, N.: Bandwidth allocation for multiservice access on EPONs. IEEE Comm. Mag. **43**(2), S16–S21 (2005). doi: 10.1109/MCOM.2005.1391498

70. Luo, Y., Ansari, N.: LSTP for dynamic bandwidth allocation and QoS provisioning over EPONs. OSA J. Opt. Netw. **4**(9), 561–572 (2005)

71. Luo, Y., Effenberger, F., Ansari, A.: Time synchronization over ethernet passive optical networks. IEEE Comm. Mag. **50**(10) (2012)

72. Luo, Y., Effenberger, F., Sui, M.: Cloud computing provisioning over passive optical networks. In: First IEEE International Conference on Communications in China (ICCC2012), Beijing, China (2012)

73. Luo, Y., Yin, S., Ansari, N., Wang, T.: Resource management for broadband access over time-division multiplexed passive optical networks. Netw. IEEE **21**(5), 20–27 (2007). doi: 10.1109/MNET.2007.4305168

74. Luo, Y., Yin, S., Wang, T., Suemura, Y., Nakamura, S., Ansari, N., Cvijetic, M.: QoS-aware scheduling over hybrid optical wireless networks. In: Optical Fiber Communication and the National Fiber Optic Engineers Conference, pp. 1–7. IEEE Conference Publications, Piscataway, NJ (2007)

75. Mandin, J.: EPON power saving via sleep mode. In: IEEE P802.3av 10GEPON Task Force Meeting, IEEE 802 LAN/MAN Standards Committee, Piscataway, NJ (2008). http://www.ieee802.org/3/av/public/2008_09/3av_0809_mandin_4.pdf

76. McGarry, M., Reisslein, M., Maier, M.: WDM Ethernet passive optical networks. IEEE Comm. Mag. **44**(2), 15–22 (2006). doi: 10.1109/MCOM.2006.1593545

77. McGarry, M.P., Reisslein, M., Colbourn, C.J., Maier, M., Aurzada, F., Scheutzow, M.: Just-in-time scheduling for multichannel EPONs. IEEE/OSA J. Lightwave Tech. **26**(10), 1204–1216 (2008)

78. McGarry, M.P., Reisslein, M., Maier, M., Keha, A.: Bandwidth management for WDM EPONs. OSA J. Opt. Netw. **5**(9), 637–654 (2006)

79. Meng, L., Assi, C., Maier, M., Dhaini, A.: Resource management in STARGATE-based Ethernet passive optical networks (SG-EPONs). IEEE/OSA J. Opt. Comm. Netw. **1**(4), 279–293 (2009)

80. Meng, L., El-Najjar, J., Alazemi, H., Assi, C.: A joint transmission grant scheduling and wavelength assignment in multichannel SG-EPON. J. Lightwave Tech. **27**(21), 4781–4792 (2009)

81. Nace, D., Pioro, M.: Max-min fairness and its applications to routing and load-balancing in communication networks: A tutorial. IEEE Comm. Surv. Tutor. **10**(4), 5–17 (2008). doi: 10.1109/SURV.2008.080403

82. Naser, H., Mouftah, H.: A joint-ONU interval-based dynamic scheduling algorithm for Ethernet passive optical networks. IEEE/ACM Trans. Netw. **14**(4), 889–899 (2006). doi: 10.1109/TNET.2006.879698

83. Oh, J.M., Koo, S.G., Lee, D., Park, S.J.: Enhancement of the performance of a reflective soa-based hybrid WDM/TDM PON system with a remotely pumped erbium-doped fiber amplifier. IEEE/OSA J. Lightwave Tech. **26**(1), 144–149 (2008)

84. Oh, S., Shin, J., Kim, K., Lee, D., Park, S., Sung, H., Baek, Y., Oh, K.: 200 GHz-spacing 8-channel multi-wavelength lasers for WDM-PON optical line terminal sources. Opt. Express **17**(11), 9401–9407 (2009)

85. Oyman, O., Foerster, J., Tcha, Y., Lee, S.: Toward enhanced mobile video services over WiMAX and LTE [WiMAX/LTE update]. IEEE Comm. Mag. **48**(8), 68–76 (2010)

86. Payne, D., Stern, J.: Transparent single-mode fiber optical networks. IEEE J. Lightwave Tech. **4**(7), 864–869 (1986)

87. Pinedo, M.: Scheduling: Theory, Algorithms, and Systems. Prentice Hall, Englewood Cliffs (2002)

88. Potts, C.: Analysis of a linear programming heuristic for scheduling unrelated parallel machines. Discrete Appl. Math. **10**, 155–164 (1985)

89. Qian, D., Cvijetic, N., Hu, J., Wang, T.: 108 Gb/s OFDMA-PON with polarization multiplexing and direct detection. IEE/OSA J. Lightwave Tech. **28**(4), 484–493 (2010)

90. Qian, D., et al.: 108 Gb/s OFDMA-PON with polarization multiplexing and direct detection. J. Lightwave Tech. **28**(4), 484–493 (2010)

91. Quax, P., Monsieurs, P., Lamotte, W., De Vleeschauwer, D., Degrande, N.: Objective and subjective evaluation of the influence of small amounts of delay and jitter on a recent first person shooter game. In: 3rd ACM SIGCOMM Workshop on Network and System Support for Games, pp. 152–156, ACM, New York (2004). doi: http://doi.acm.org/10.1145/1016540.1016557

92. Raake, A.: Short-and long-term packet loss behavior: towards speech quality prediction for arbitrary loss distributions. IEEE Trans. Audio Speech Lang. Process. **14**(6), 1957–1968 (2006)

93. Radunovic, B., Le Boudec, J.Y.: A unified framework for max-min and min-max fairness with applications. IEEE/ACM Trans. Netw. **15**(5), 1073–1083 (2007). doi: 10.1109/TNET.2007.896231

94. Ramantas, K., Vlachos, K., Ellinas, G., Hadjiantonis, A.: Efficient resource management via dynamic bandwidth sharing in a WDM-PON ring-based architecture. In: 14th International Conference on Transparent Optical Networks (ICTON). IEEE Conference Publications, Piscataway, NJ (2012)

95. Reaz, A., Ramamurthi, V., Tornatore, M.: Cloud-over-WOBAN (CoW): An offloading-enabled access network design. In: IEEE International Conference on Communications (ICC). IEEE Conference Publications, Piscataway, NJ (2011)

96. Reaz, A., Ramamurthi, V., Tornatore, M., Mukherjee, B.: Green provisioning of cloud services over wireless-optical broadband access networks. In: Proceedings of the IEEE Globecom. IEEE Conference Publications, Piscataway, NJ (2011)

97. Reichl, P., Tuffin, B., Schatz, R.: Economics of logarithmic quality-of-experience in communication networks. In: The 9th Conference on Telecommunications Internet and Media Techno Economics (CTTE), pp. 1–8, IEEE Conference Publications, Piscataway, NJ (2010). doi: 10.1109/CTTE.2010.5557702

98. Sanneck, H., Carle, G., Koodli, R.: Framework model for packet loss metrics based on loss runlengths. In: SPIE/ACM SIGMM Multimedia Computing and Networking, ACM, New York (2000)

99. Sarkar, S., Yen, H., Dixit, S., Mukherjee, B.: A mixed integer programming model for optimum placement of base stations and optical network units in a hybrid wireless-optical broadband access network (WOBAN). In: IEEE Wireless Communications and Networking Conference (WCNC), pp. 3907–3911, IEEE, Piscataway, NJ (2007)

100. Sarkar, S., Yen, H., Dixit, S., Mukherjee, B.: Hybrid wireless-optical broadband access network (WOBAN): Network planning using Lagrangean relaxation. IEEE/ACM Trans. Netw. **17**(4), 1094–1105 (2009)

101. Shaw, W.T., Wong, S.W., Cheng, N., Balasubramanian, K., Zhu, X., Maier, M., Kazovsky, L.: Hybrid architecture and integrated routing in a scalable optical wireless access network. IEEE/OSA J. Lightwave Tech. **25**(11), 3443–3451, IEEE Conference Publications, Piscataway, NJ (2007). doi: 10.1109/JLT.2007.909202

102. Shchepin, E.V., Vakhania, N.: An optimal rounding gives a better approximation for scheduling unrelated machines. Oper. Res. Lett. **33**, 127–133 (2005)

103. Song, H., Kim, B., Mukherjee, B.: Multi-thread polling: A dynamic bandwidth distribution scheme in long-reach PON. IEEE J. Sel. Areas Comm. **27**(2), 134–142 (2009)

104. Tanaka, K.: 10G-EPON standardization and its development status. In: Optical Fiber Communication (OFC), pp. 1–20, 22–26, Optical Society of America 2010 Massachusetts Ave., N.W. Washington, D.C. 20036-1012 USA (2009)

105. Tasaka, S., Ishibashi, Y.: Mutually compensatory property of multimedia QoS. IEEE Int. Conf. Comm. **2**, 1105–1111 (2002). doi: 10.1109/ICC.2002.997023

106. The carrier cloud. Strategic white paper www.alcatel-lucent.com (2011). Accessed on 2011

107. Tongia, R.: Can broadband over powerline carrier (PLC) compete? A techno-economic analysis. Telecomm. Policy **28**(7–8), 559–578 (2004)

108. Vadgama, S.: Trends in green wireless access. Fujitsu Sci. Tech. J. **45**(4), 404–408 (2009)

109. Verma, D., Zhang, H., Ferrari, D.: Delay jitter control for real-time communication in a packet switching network. In: IEEE TRICOMM, pp. 35–43, IEEE Conference Publications, Piscataway, NJ (1991). doi: 10.1109/TRICOM.1991.152873

110. Wang, C., Wei, W., Zhang, W., Jiang, H., Qiao, C., Wang, T.: Optimal wavelength scheduling for hybrid WDM/TDM passive optical networks. IEEE/OSA J. Opt. Comm. Netw. **3**(6), 522–532 (2011)

111. Wei, W., Wang, T., Qian, D., Hu, J.: MAC protocols for optical orthogonal frequency division multiple access (OFDMA)-based passive optical networks. In: Optical Fiber communication/National Fiber Optic Engineers Conference, OFC/NFOEC. Conference on, pp.1–3, 24–28, Optical Society of America 2010 Massachusetts Ave., N.W. Washington, D.C. 20036–1012 USA (2008). doi: 10.1109/OFC.2008.4528240

112. Weinstein, S., Luo, Y., Wang, T.: Passive Optical Networks. IEEE Conference Publications, Piscataway, NJ (2012)

113. Wong, S-W., Valcarenghi, L., Yen, S-H., Campelo, D.R., Yamashita, S., Kazovsky, L.: Sleep Mode for Energy Saving PONs: Advantages and Drawbacks. In: IEEE GLOBECOM Workshops, pp. 1–6, IEEE Conference Publications, Piscataway, NJ (2009). doi: 10.1109/GLOCOMW.2009.5360736

114. Xue, D., Qin, Y., Siew, C.K.: Deterministic qos provisioning with network calculus based admission control in WDM EPON networks. In: IEEE International Conference on Communications, pp. 1–6, IEEE Conference Publications, Piscataway, NJ (2009). doi: 10.1109/ICC.2009.5198914

115. Yan, B., Guo, W., Jin, Y., Hu, W.: A novel dynamic wavelength bandwidth allocation scheme over OFDMA PONs. In: Asia Communications and Photonics Conference (ACP), vol. 8310, p. 17, IEEE Conference Publications, Piscataway, NJ (2011)

116. Yan, Y., Dittmann, L.: Energy efficiency in ethernet passive optical networks (EPONs): Protocol design and performance evaluation. J. Comm. **6**(3), 249–261 (2011)

117. Yan, Y., Wong, S., Valcarenghi, L., Yen, S., Campelo, D., Yamashita, S., Kazovsky, L., Dittmann, L.: Energy management mechanism for ethernet passive optical networks (EPONs). In: IEEE International Conference on Communications (ICC). IEEE Conference Publications, Piscataway, NJ (2010)

118. Yang, C.: Code space enlargement for hybrid fiber radio and baseband OCDMA PONs. J. Lightwave Tech. **29**(9), 1394–1400 (2011)

119. Yeh, J., Chen, J., Lee, C.: Comparative Analysis of Energy-Saving Techniques in 3GPP and 3GPP2 Systems. IEEE Trans. Vehicular Tech. **58**(1), 432–448 (2009)

120. Yin, S., Ansari, N.: Nonlinear predictor-based dynamic resource allocation over point-to-multipoint (p2mp) networks: A control theoretical approach. IEEE/OSA J. Opt. Comm. Netw. **2**(12), 1052–1062 (2010). doi: 10.1364/JOCN.2.001052

121. Yoshino, M., Miki, N., Yoshimoto, N., Kumozaki, K.: Multiwavelength optical source for OCDM using sinusoidally modulated laser diode. IEEE/OSA J. Lightwave Tech. **27**, 4524–4529 (2009)

122. Yu, W., Ginis, G., Cioffi, J.: Distributed multiuser power control for digital subscriber lines. IEEE J. Sel. Areas Comm. **20**(5), 1105–1115 (2002)

123. Zhang, J., Ansari, N.: Utility max-min fair resource allocation for diversified applications in EPON. In: AccessNets, Hongkong, China (2009)

124. Zhang, J., Ansari, N.: An application-oriented resource allocation scheme for EPON. IEEE Syst. J. **4**(4), 424–431 (2010)

125. Zhang, J., Ansari, N.: Design of WDM PON with tunable lasers: The upstream scenario. IEEE/OSA J. Lightwave Tech. **28**(2), 228–236 (2010)

126. Zhang, J., Ansari, N.: Dynamic time allocation and wavelength assignment in next-generation multi-rate multi-wavelength passive optical networks. In: Proceedings of the IEEE ICC, Cape Town, South Africa (2010)

127. Zhang, J., Ansari, N.: On assuring end-to-end QoE in next generation networks: Challenges and a possible solution. IEEE Comm. Mag. **49**(7), 185–191 (2011)

128. Zhang, J., Ansari, N.: On OFDMA resource allocation and wavelength assignment in OFDMA-based WDM radio-over-fiber picocellular systems. IEEE J. Sel. Areas Comm. **29**(6), 1273–1283 (2011) [Special Issue on Distributed Broadband Wireless Communications]

129. Zhang, J., Ansari, N.: On the capacity of WDM passive optical networks. IEEE Trans. Comm. **59**(2), 552–559 (2011)

130. Zhang, J., Ansari, N.: Scheduling hybrid WDM/TDM passive optical networks with non-zero laser tuning time. IEEE/ACM Trans. Netw. **19**(4), 1014–1027 (2011)

131. Zhang, J., Ansari, N.: Towards energy-efficient 1G-EPON and 10G-EPON with sleep-aware MAC control and scheduling. IEEE Comm. Mag. **49**(2), S33–S38 (2011)

132. Zhang, J., Ansari, N.: Extending onu lifetime beyond 72 hours in EPON for emergency communications. In: 2012 International Conference on Computing, Networking and Communications (ICNC), pp. 287–291. IEEE Conference Publications, Piscataway, NJ (2012)

133. Zhang, J., Ansari, N.: On preemptive multi-wavelength scheduling in hybrid WDM/TDM passive optical networks. IEEE/OSA J. Opt. Comm. Netw. **4**, 238–247 (2012)

134. Zhang, J., Ansari, N.: Standards-compliant EPON sleep control for energy efficiency: Design and analysis, Communications (ICC), pp. 2994–2998, 10–15, IEEE Conference Publication, Piscataway, NJ (2012). doi: 10.1109/ICC.2012.6364591

135. Zhang, J., Ansari, N., Luo, Y., Effenberger, F., Ye, F.: Next-generation PONs: A performance investigation of candidate architectures for next-generation access stage 1. IEEE Comm. Mag. **47**(8), 49–57 (2009). doi: 10.1109/MCOM.2009.5181892

136. Zhang, J., Wang, T., Ansari, N.: Designing energy-efficient optical line terminal for TDM passive optical networks. In: 34th IEEE Sarnoff Symposium. IEEE Conference Publications, Piscataway, NJ (2011)

137. Zhang, J., Wang, T., Ansari, N.: An efficient MAC protocol for asynchronous ONUs in OFDMA PONs. In: Proceedings of Optical Fiber Communication Conference and Exposition (OFC), Optical Society of America 2010 Massachusetts Ave., N.W. Washington, D.C. 20036–1012, USA (2011)

138. Zheng, J., Mouftah, H.: Media access control for Ethernet passive optical networks: An overview. IEEE Comm. Mag. **43**(2), 145–150 (2005). doi: 10.1109/MCOM.2005.1391515